Introduction to
THERMAL ANALYSIS

Techniques and applications

Introduction to
THERMAL ANALYSIS

Techniques and applications

Michael E. Brown

Department of Chemistry and Biochemistry
Rhodes University, Grahamstown, 6140
South Africa

London
CHAPMAN AND HALL
New York

First published in 1988 by
Chapman and Hall Ltd
11 New Fetter Lane, London EC4P 4EE
Published in the USA by
Chapman and Hall
29 West 35th Street, New York NY 10001

© *1988, Michael E. Brown*

Printed in Great Britain at the
University Press, Cambridge

ISBN 0 412 30230 6

British Library Cataloguing in Publication Data

Brown, Michael E.
   Introduction to thermal analysis:
   techniques and applications.
   1. Thermal analysis
   I. Title
   543'.086   QD79.T38

      ISBN 0-412-30230-6

Library of Congress Cataloging in Publication Data

Brown, Michael E., 1938–
   Introduction to thermal analysis: techniques and applications/
   Michael E. Brown.
      p.   cm.
   Includes bibliographies and index.
   ISBN 0-412-30230-6
   1. Thermal analysis.   I. Title.
QD79.T38B76 1988
543'.086—dc19

Contents

Preface

The aim of this book is, as its title suggests, to help someone with little or no knowledge of what thermal analysis can do, to find out briefly what the subject is all about, to decide whether it will be of use to him or her, and to help in getting started on the more common techniques. Some of the less-common techniques are mentioned, but more specialized texts should be consulted before venturing into these areas.

This book arose out of a set of notes prepared for courses on thermal analysis given at instrument workshops organized by the S.A. Chemical Institute. It has also been useful for similar short courses given at various universities and technikons. I have made extensive use of the manufacturers' literature, and I am grateful to them for this information. A wide variety of applications has been drawn from the literature to use as examples and these are acknowledged in the text. A fuller list of the books, reviews and other literature of thermal analysis is given towards the back of this book. The ICTA booklet 'For Better Thermal Analysis' is also a valuable source of information.

I am particularly grateful to my wife, Cindy, for typing the manuscript, to Mrs Heather Wilson for the line drawings, and to Professor David Dollimore of the University of Toledo, Ohio, for many helpful suggestions.

Michael E. Brown
Grahamstown 1987

Thermal analysis

is the measurement of changes in physical properties of a substance as a function of *temperature* whilst the substance is subjected to a controlled temperature programme

Chapter 1

Introduction

1.1 Definition and history

What do bread and chocolate, hair and finger-nail clippings, coal and rubber, ointments and suppositories, explosives, kidney stones and ancient Egyptian papyri have in common? Many interesting answers could probably be suggested, but the connection wanted in this context is that they all undergo interesting and practically important changes on heating.

The study of the effect of heat on materials obviously has a long history, from man's earliest attempts at producing pottery, extracting metals ($\sim$8000 BC) and making glass ($\sim$3400 BC) through the philosophical discussions of the alchemists on the elements of fire, air, earth and water, to early work on the assaying of minerals ($\sim$1500 AD), followed by the development of thermometry and calorimetry [1, 2]. Only in the late 19th century did experiments on the effect of heat on materials become more controlled and more quantitative. Much of this work depended upon the development of the analytical balance which has its own interesting history [3, 4].

The establishment of the International Confederation of Thermal Analysis (ICTA) in Aberdeen in 1965 and the great advances in commercially available equipment have resulted in thermal analysis being an extremely active field with applications in numerous directions.

There has been some discussion [5, 6] on whether thermal analysis can really be regarded as an analytical method. It was concluded [5] that thermal analysis could be fitted into the general systems theory model of chemical analysis, based on three 'elements': the sample, the reagent and the signal, by considering heat as a reagent (Fig. 1.1(a)).

It has also been suggested [5] that the definition of thermal analysis should be extended to allow for rapid heating of the sample to some elevated temperature, followed by measurement of the property with time under isothermal conditions, i.e. at zero heating rate.

Introduction

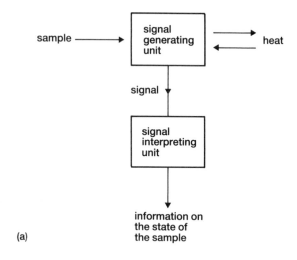

(a)

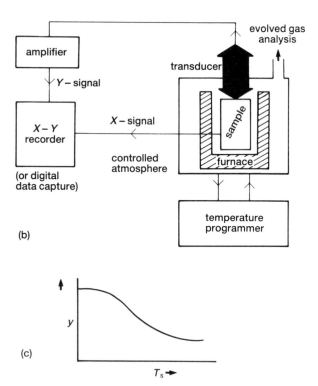

(b)

(c)

Figure 1.1 (a) Systems theory model of thermal analysis. (b) Generalized thermal analysis instrument. (c) Thermal analysis curve (note: 'thermogram' is not an approved term). (Based on Levy, P. F. (1971) *Int. Lab.*, Jan/Feb, 61, with the permission of International Scientific Communications, Inc.)

1.2 Thermal analysis instruments

All thermal analysis instruments have features in common. These are illustrated in Fig. 1.1(b). The variety of the techniques to be discussed stems from the variety of physical properties that can be measured and the variety of transducers that can be used to convert these properties into electrical signals.

Measurements are usually continuous and the heating rate is often, but not necessarily, linear with time. The result of such measurements is a **thermal analysis curve** (Fig. 1.1(c)) and the features of this curve (peaks, discontinuities, changes of slope, etc.) are related to **thermal events** in the sample. The thermal events which may be detected are described in chapter 2.

1.3 Types of measurement

Absolute values of the sample property may be recorded, or the difference in the property of the sample compared to the same property of a reference material may be more convenient to measure, or the rate of change of the sample property with temperature may be of interest (derivative measurements).

1.4 The main techniques

The main sample properties used and the associated techniques are listed in Table 1.1. All of these techniques may be combined with evolved gas analysis

Table 1.1 The main thermal analysis techniques

Property	*Technique*	*Abbreviation*
Mass	Thermogravimetry*	TG
	Derivative thermogravimetry	DTG
Temperature	Differential thermal analysis	DTA
Enthalpy	Differential scanning calorimetry	DSC
Dimensions	Thermodilatometry	
Mechanical properties	Thermomechanical analysis (Thermomechanometry)	TMA
	Dynamic mechanical analysis	DMA
Optical properties	Thermoptometry or thermomicroscopy	
Magnetic properties	Thermomagnetometry	TM
Electrical properties	Thermoelectrometry	
Acoustic properties	Thermosonimetry and thermoacoustimetry	TS
Evolution of radioactive gas	Emanation thermal analysis	ETA
Evolution of particles	Thermoparticulate analysis	TPA

*The term 'thermogravimetric analysis' and the abbreviation 'TGA' are in common use but are not approved by the ICTA Nomenclature Committee.

(EGA) (see chapter 10), and it is often possible to carry out simultaneous measurements of more than one property (see chapter 9).

1.5 References

1. Mackenzie, R. C. (1984) *Thermochim. Acta*, **73**, 251, 307; (1985), **92**, 3.
2. Szabadvary, F. and Buzagh-Gere, E. (1979) *J. Thermal Anal.*, **15**, 389.
3. Keattch, C. J. and Dollimore, D. (1975) *An Introduction to Thermogravimetry*, Heyden, London, 2nd edn.
4. Vieweg, R. (1972) in *Progress in Vacuum Microbalance Techniques*, Vol. 1, (eds. T. Gast and E. Robens), Heyden, London, p. 1.
5. Meisel, T. (1984) *J. Thermal Anal.*, **29**, 1379.
6. Sestak, J. (1979) *Thermochim. Acta*, **28**, 127.

Chapter 2

Thermal events

2.1 Reactions of solids

When a single pure solid substance, A, is heated in an inert atmosphere, the resultant increase in molecular, atomic or ionic motion may lead to changes in crystal structure, sintering, melting or sublimation [1, 2]. If the intramolecular forces are weaker than the intermolecular forces, the substance may decompose forming new molecular fragments, some or all of which may be volatile at the temperatures reached. Numerous examples of such decompositions exist [3, 4], e.g.

$$BaCl_2 . 2H_2O(s) \rightarrow BaCl_2(s) + 2H_2O(g)$$

$$CaCO_3(s) \rightarrow CaO(s) + CO_2(g)$$

$$NH_4Cl(s) \rightarrow NH_3(g) + HCl(g)$$

$$BaN_6(s) \rightarrow Ba(s) + 3N_2(g)$$

More complicated reactions result when the initial solid can react with the surrounding atmosphere [5], e.g.

$$2Ag(s) + \tfrac{1}{2}O_2(g) \rightarrow Ag_2O(s)$$

$$CuO(s) + H_2(g) \rightarrow Cu(s) + H_2O(g)$$

$$Ni(s) + 4CO(g) \rightarrow Ni(CO)_4(g)$$

$$C(s) + O_2(g) \rightarrow CO_2(g)$$

When more than one solid substance is present initially, there are correspondingly more possibilities for interaction on heating and new phases, such as solid solutions and eutectic mixtures, may form as well as new compounds formed by addition or double decomposition reactions [6], e.g.

$$Fe_2O_3(s) + MgO(s) \rightarrow MgFe_2O_4(s)$$

$$NaCl(s) + AgBr(s) \rightarrow AgCl(s) + NaBr(s)$$

The above changes are nearly always accompanied by enthalpy changes, and sometimes also by changes in mass, so they may be studied using one or more of the thermal analysis techniques listed in Table 1.1. The main thermal events are summarized in Table 2.1.

Table 2.1 Thermal events

$A(s_1) \longrightarrow$	$A(s_2)$	phase transition
	$A(l)$	melting
	$A(g)$	sublimation
	$B(s) + gases$ $\Big\}$ decomposition $\Big\{$ thermal	
	$gases$	radiolytic
$A(glass) \rightarrow A(rubber)$		glass transition
$A(s) + B(g) \rightarrow C(s)$ $\Big\{$		oxidation / tarnishing
$A(s) + B(g) \rightarrow gases$ $\Big\{$		combustion / volatilization
$A(s) + (gases)_1 \rightarrow A(s) + (gases)_2$		heterogeneous catalysis
$A(s) + B(s) \rightarrow AB(s)$		addition
$AB(s) + CD(s) \rightarrow AD(s) + CB(s)$		double decomposition

2.2 References

1. Rao, C. N. R. and Rao, K. J. (1978) *Phase Transitions in Solids*, McGraw-Hill, New York.
2. West, A. R. (1984) *Solid State Chemistry*, Wiley, Chichester, Ch. 12.
3. Galwey, A. K. (1967) *Chemistry of Solids*, Chapman and Hall, London.
4. Brown, M. E., Dollimore, D. and Galwey, A. K. (1980) *Reactions in the Solid State, Comprehensive Chemical Kinetics*, Vol. 22, Elsevier, Amsterdam.
5. Szekely, J., Evans, J. W. and Sohn, H. Y. (1976) *Gas-Solid Reactions*, Academic Press, New York.
6. Schmalzried, H. (1981) *Solid State Reactions*, Verlag Chemie, Weinheim, 2nd edn.

Chapter 3

Thermogravimetry (TG)

3.1 Introduction

Measurements of changes in sample mass with temperature are made using a thermobalance. This is a combination of a suitable electronic microbalance with a furnace and associated temperature programmer. The balance should be in a suitably enclosed system so that the atmosphere can be controlled (Fig. 3.1 and section 3.4).

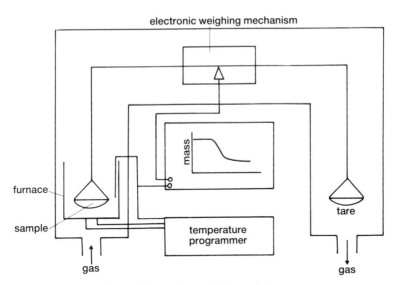

Figure 3.1 A schematic thermobalance.

3.2 The balance

Several types of balance mechanism are possible [1–3]. These include beam, spring, cantilever and torsion balances. Some operate on measurements of deflection, while others operate in null mode. Null-point weighing mechanisms are favoured in TG as they ensure that the sample remains in the same zone of the furnace irrespective of changes in mass.

Various sensors have been used to detect deviations of the balance beam from the null-position, e.g. the Cahn RG electrobalance (Fig. 3.2) uses an electro-optical device with a shutter attached to the balance beam. The shutter partly blocks the light path between a lamp and a photocell. Movement of the beam alters the light intensity on the photocell and the amplified output from the photocell is used to restore the balance to its null-point and, at the same time, is a measure of the mass change. The restoring mechanism is electromagnetic. The beam has a ribbon suspension and a small coil at the fulcrum, located in the field of a permanent magnet. The coil exerts a restoring force on the beam proportional to the current from the photocell.

Provision is also usually made for electrical taring and for scale expansion to give an output of mass loss as a percentage of the original sample mass.

The sensitivity of a thermobalance and the maximum load which it can accept, without damage, are related. Typical values are maximum loads of 1 g and sensitivities of the order of 1 μg. Use can be made of alternative suspension points on the balance beam to increase the maximum load.

The output signal may be differentiated electronically to give a derivative thermogravimetric (DTG) curve.

In addition to conventional TG studies, much use has been made of microbalances in studying the kinetics of reactions of solids (see chapter 13) under isothermal conditions [4]. Studies of decomposition reactions pose some special problems. Many decompositions are reversible if a supply of the gaseous products of reaction is maintained. Such studies thus require careful control of the surrounding atmosphere and should include runs at reduced pressures. Reducing the pressure, though, worsens the heat exchange, causing problems in temperature measurement. In an attempt to overcome some of these problems, the use of quartz crystal microbalances has been suggested.

Use of the piezoelectric effect in certain crystals (usually quartz) for measuring the mass of material deposited or condensed on a crystal face is well documented [5, 6]. There are basically two ways in which such crystals can be used in TG studies. The sample may be deposited on a crystal face and the sample plus crystal may be heated as required, or the sample may be heated separately in one part of a reaction chamber and all or some of its gaseous decomposition products may be condensed on the initially clean face of a crystal which is held at a suitably low temperature. Changes in the amount of material deposited on the crystal surface show up as changes of frequency of oscillation of the crystal, which is usually excited in a conventional series resonance circuit. The observed change in frequency depends on the value of the frequency itself and the mass and area of the coating on the crystal face. Mass changes of as little as 10^{-12} g can be detected.

Quartz crystals also show marked changes in resonance frequency with

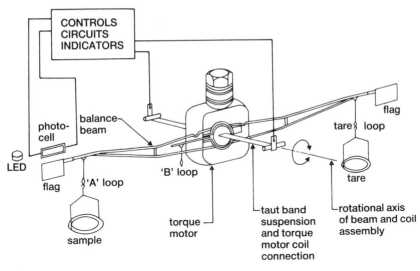

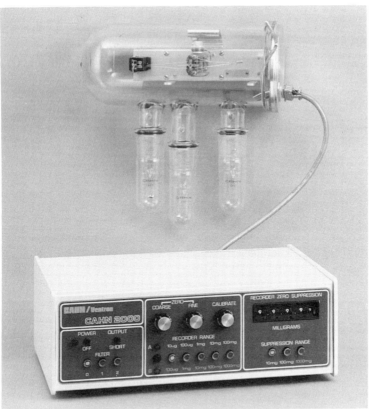

Figure 3.2 Electronic microbalance with null-mechanism. (With the permission of Cahn Instruments, Inc.)

temperature, so most studies have been limited to isothermal conditions. The maximum temperature at which quartz exhibits pyroelectric behaviour is 573°C. Lithium niobate crystals can be used to extend this range to beyond 1000°C.

Henderson *et al.* [7] have used a quartz crystal for non-isothermal TG by using a computer to record and correct for the frequency–temperature and the mass–frequency relationships of the crystal to obtain the TG (mass–temperature) curve. Descriptions of the TG system, including arrangements for heating the crystal plus sample, and of the hardware and software required are given [7]. The system was used to record TG curves of several coordination compounds. Samples ($\sim 1 \mu g$) were applied to the quartz crystal face as solutions in dichloromethane.

The less direct approach, that of determining the mass of gaseous decomposition products, condensed on a cooled quartz crystal [8, 9] avoids the frequency–temperature relationship problem and also removes some of the problems of heating and measuring the temperature of the sample. The sample can be heated in a conventional furnace (with contact temperature measurement) in a system where the pressure can be reduced as desired. Usually the difference in frequency between the 'weighing' crystal and a matched reference crystal (in a separate chamber) is amplified and recorded. The temperature of the weighing crystal is also recorded. For quantitative studies the system has to be calibrated for different gaseous products and sample geometry.

A microbalance based on the mass dependence of the resonance frequency of a vibrating quartz wire, from which loads are suspended, has been described by Mulder [10].

3.3 Heating the sample

In the more conventional thermobalances, there are three main variations in the position of the sample relative to the furnace as shown in Fig. 3.3. The furnace is normally an electrical resistive heater and may also, as shown, be within the balance housing, part of the housing, or external to the housing. It should: (i) be non-inductively wound, (ii) be capable of reaching 100 to 200°C above the maximum desired working temperature, (iii) have a uniform hot-zone of reasonable length, (iv) reach the required starting temperature as quickly as possible (i.e. have a low heat capacity), and (v) not affect the balance mechanism through radiation or convection. Transfer of heat to the balance mechanism should be minimized by inclusion of radiation shields and convection baffles.

Furnaces containing electrical resistive heaters [11] rely mainly on heating of the sample by conduction, through solid or gas, with inevitable large temperature gradients, especially when dealing with samples of low thermal conductivity such as polymers and inorganic glasses. Heating by radiation

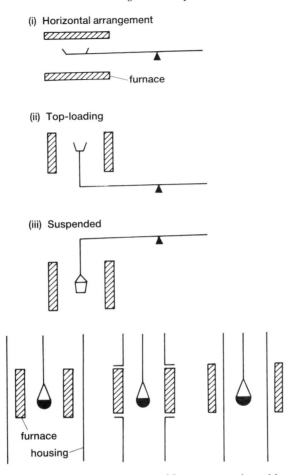

(i) Horizontal arrangement

furnace

(ii) Top-loading

(iii) Suspended

furnace
housing

Figure 3.3 Alternative arrangements of furnace, sample and housing.

becomes significant only at high temperatures in such furnaces, but alternative heating systems using either infrared or microwave radiation have been considered.

For infrared heating [12, 13] the light from several halogen lamps is focused onto the sample by means of elliptic [12] or parabolic [13] reflectors. Temperatures of over 1400°C may be achieved at heating rates of up to 1000°C min^{-1} and control within ± 0.5°C.

Karmazin *et al.* [14, 15] have suggested the use of microwaves to generate heat uniformly within the sample. Such a process would have the immediate advantage of allowing for the use of larger and more representative samples than are often used in thermal analysis. Temperature measurement and power control both present problems in using microwaves. A thermocouple may be positioned in the waveguide if its orientation is such that it collects no energy itself from the microwaves, and its size is small compared to the guide so that reflected waves are negligible. The thermocouple is then used [14] for power control *via* a microcomputer system.

For DTA (chapter 4) the sample and the reference are placed in cylindrical Teflon or silica containers. The absorption coefficient for microwaves varies from one substance to another and also varies markedly with temperature, thus making matching of the sample with a suitable reference difficult. As with classical DTA, the sample is often mixed with an excess of inert reference material, such as Al_2O_3 or TiO_2 which have high absorption coefficients. The higher the absorption coefficient, the greater the heating rate possible.

Hydrates show interesting behaviour on heating with microwaves since the absorption coefficient for H_2O is so high (e.g. ~ 1000 times that of $CaSO_4$ in gypsum). Conventional DTA shows dehydration as an endothermic process, but with microwave heating the process appears to be exothermic.

Similarly, it was found [15] that in a microwave-heated thermodilatometer (chapter 6) the results differed considerably from those obtained using conventional heating.

The possibility has also been suggested [16] of the use of lasers for heating and infrared pyrometers for remote temperature measurement, in the *in situ* thermal analysis of bulk materials.

Steinheil [17] described a vacuum microbalance system, for use at high temperatures (1600 to 2400°C) which incorporated induction heating by a high-frequency electromagnetic field, but did not allow weighing during actual heating because of the electromagnetic forces acting on the sample.

3.4 The atmosphere

Thermobalances are normally housed in glass or metal systems [18] (Figs 3.1 and 3.4), to allow for operation at pressures ranging from high vacuum ($< 10^{-4}$ Pa) to high pressure (> 3000 kPa), of inert, oxidizing, reducing or corrosive gases. Microbalances are affected by a variety of disturbances [1, 18]. One correction which may have to be made is for buoyancy arising from lack of symmetry in the weighing system. If the asymmetry can be measured in terms of a volume ΔV, the mass of displaced gas (assuming ideal gas behaviour) is $m = pM\Delta V/RT$ (where p is the pressure and M the molar mass). The buoyancy thus depends not only on the asymmetry ΔV, but also on the pressure, temperature and nature of the gas. Attempts may be made to reduce ΔV to zero, or a correction may be calculated, or an empirical correction may be applied by heating an inert sample under similar conditions to those to be used in the study of the sample of interest.

At low pressures (10^{-2} to 270 Pa) [1], a particular problem is thermomolecu-

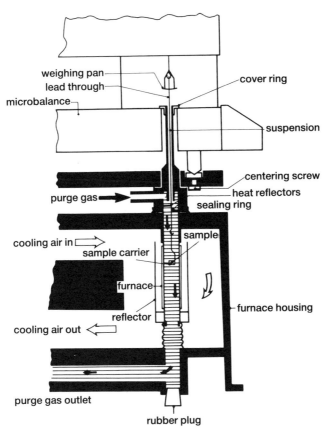

Figure 3.4 Mettler TA3000 – The TG50 Thermobalance. (With the permission of Mettler Instrumente AG.)

lar flow [1] which results when there is a temperature gradient along the sample holder and support. This gradient causes 'streaming' of molecules in the direction hot→cold, i.e. up the suspension as a rule, giving spurious mass changes. Thermomolecular flow may be minimized (i) by working outside the pressure range by adding inert gas, or (ii) by careful furnace design and sample placement, including use of a symmetrical balance design with twin furnaces, or (iii) by determination of corrections required, using an inert sample, as is done for buoyancy corrections.

The sample may be heated in a small container with a restricted opening. Decomposition then occurs in a self-generated atmosphere [19] of gaseous decomposition products. The inhibiting or catalytic effects of these products on the decomposition can then be studied.

Garn and co-workers [20, 21] have described the design and use of furnaces for carrying out reactions in controlled atmospheres of ligand. Such systems are essential for studying the thermodynamic reversibility of dissociations of coordination compounds.

At atmospheric pressure, the atmosphere can be static or flowing. A flowing atmosphere has the advantages that it: (i) reduces condensation of reaction products on cooler parts of the weighing mechanism; (ii) flushes out corrosive products; (iii) reduces secondary reactions; and (iv) acts as a coolant for the balance mechanism. The balance mechanism should, however, not be disturbed by the gas flow. It is possible for the balance mechanism to be protected by an inert gas atmosphere, while a corrosive gas or vapour is passed over the sample (e.g. H_2O vapour for dehydration studies).

The noise level of TG traces at pressures above ~ 20 kPa usually increases as the temperature increases on account of thermal convection [1, 22, 23]. The use of dense carrier gases at high pressures in hot zones with large temperature gradients gives the most noise. Variations in the flow rate of the gas do not affect the noise level much, but may shift the weighing zero. Noise levels also increase as the radius of the hangdown tube increases. Thermal convection, and hence noise, can be reduced by altering the gas density gradient by introducing a low density gas, such as helium, above the hot region. Alternatively, and more practically, baffles can be introduced in the hangdown tube. A series of close-fitting convoluted baffles (Fig. 3.5) was found to be most successful [24]. Even with the baffle it was found that changes in ambient temperature caused non-turbulent gas movements in the hangdown tube and hence apparent changes of mass.

Figure 3.5 Reduction of convection effects by use of baffles in the hangdown tube.

Sample containers should be of low mass and made of inert material (e.g. platinum containers may catalyze some reactions). Samples should generally be thinly spread to allow for ready removal of evolved gases. Results should be checked for sample-holder geometry effects.

Build up of electrostatic charge on glass housings is another source of spurious mass effects. Methods proposed for dealing with this problem include use of weak radioactive sources for ionization of the balance atmosphere, coating of glassware with a sputtered metal film or other metal shielding, and the use of various commercial sprays or solutions. Wiping the outside of the glassware with ethanol is reasonably effective.

3.5 The sample

Although solid samples may be nominally of the same chemical composition, there may be considerable differences in their behaviour on heating. These differences arise from structural differences in the solid, such as the defect content, the porosity and the surface properties, which are dependent on the way in which the sample is prepared and treated after preparation. For example, very different behaviour will generally be observed for single crystals compared to finely ground powders of the same compound [25]. In addition to the influence of defects on reactivity [26], the thermal properties of powders differ markedly from those of the bulk material.

As the amount of sample used increases, several problems arise. The temperature of the sample becomes non-uniform through slow heat transfer and through either self-heating or self-cooling as reaction occurs. Also exchange of gas with the surrounding atmosphere is reduced. These factors may lead to irreproducibility. Even when the sample material is inhomogeneous and hence a larger sample becomes desirable (e.g. coal and mineral samples), the sample mass should be kept to a minimum and replicates examined for reproducibility if necessary. Small sample masses also protect the apparatus in the event of explosion or deflagration. The sample should be powdered where possible and spread thinly and uniformly in the container [25].

3.6 Temperature measurement and calibration

The sample temperature, T_s, will usually lag behind the furnace temperature, T_f, and T_s cannot be measured very readily without interfering with the weighing process. The lag, $T_f - T_s$, may be as much as 30°C, depending upon the operating conditions. The lag is marked when operating in vacuum or in fast-flowing atmospheres and with high heating rates. Temperature measurement is usually by thermocouple and it is advisable to have separate thermocouples for

measurement of T_s and for furnace regulation. (Platinum resistance thermometers are used in some controllers.)

An ingenious method of temperature calibration [27] for small furnaces
makes use of the Curie points of a range of metals and alloys. On heating a
ferromagnetic material, it loses its ferromagnetism at a characteristic temperature known as the Curie point. If a magnet is positioned below the sample of
ferromagnetic material, as shown in Fig. 3.6(a), the total downward force on the
sample, at temperatures below the Curie point, is the sum of the sample weight
and the magnetic force. At the Curie point the magnetic force is reduced to zero
and an apparent mass loss is observed (Fig. 3.6(b)). By using several
ferromagnetic materials, a multi-point temperature calibration may be
obtained (Fig. 3.6(c)). It has also been suggested that inert mass-pieces be
suspended from the balance, using links of fusible wire so that disturbances will
be caused as the fusion temperature is reached [31–33].

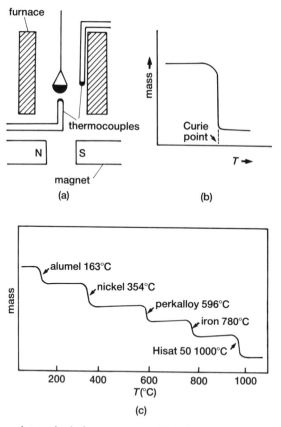

Figure 3.6 Curie-point method of temperature calibration [27]. (With the permission of
Thermochimica Acta, Elsevier, Amsterdam.)

Even with careful temperature calibration, T_s may still not be accurately known, as slow heat transfer may cause self-heating or self-cooling from strongly exothermic or endothermic processes in relatively large samples.

3.7 Temperature programmers

A wide variety of fully-electronic temperature controllers and programmers is available commercially. Temperature sensors are either platinum resistance thermometers or thermocouples. Heating rates offered range from fractions of a degree per minute to nearly 1000°C min^{-1}, with facilities for isothermal operation. In a small thermal analysis laboratory, one programmer may serve several instruments [28].

3.8 Interpretation of TG and DTG curves

Actual TG curves obtained may be classified into various types [29] as illustrated in Fig. 3.7. Possible interpretations are as follows.

Type (i) curves. The sample undergoes no decomposition with loss of volatile products over the temperature range shown. No information is obtained, however, on whether solid phase transitions, melting, polymerization or other reactions involving no volatile products have occurred. Use of some of the other techniques (Table 1.1) is necessary to eliminate these possibilities. Assuming that these possibilities are eliminated, the sample would then be known to be stable over the temperature range considered. This could be good news if a heat resistant material was being sought, or bad news if potential explosives were being tested!

Type (ii) curves. The rapid initial mass loss observed is characteristic of desorption or drying. It could also arise, when working at reduced pressures, from effects such as thermomolecular flow or convection (section 3.4). To check that the mass loss is real, it is advisable to rerun the sample, which should then produce a type (i) curve, unless the carrier gas contained moisture or was very readily readsorbed on the sample at the lower temperature.

Type (iii) curves. These represent decomposition of the sample in a single stage. The curve may be used to define the limits of stability of the reactant, to determine the stoichiometry of the reaction, and to investigate the kinetics of reaction (see chapter 13).

Type (iv) curves. These indicate multi-stage decomposition with relatively stable intermediates. Again, the temperature limits of stability of the reactant and of the intermediates can be determined from the curve, together with the more complicated stoichiometry of reaction.

Type (v) curves. These also represent multi-stage decomposition, but in this

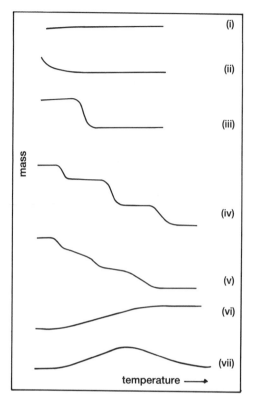

Figure 3.7 Main types of thermogravimetric (TG) curves. (Based on Duval, C. (1963) *Inorganic Thermogravimetric Analysis*, Elsevier, Amsterdam, 2nd edn. and Daniels, T. (1973) *Thermal Analysis*, Kogan Page, London, with permission.)

example stable intermediates are not formed and little information on all but the stoichiometry of the overall reaction can be obtained. It is important to check the effect of heating rate on the resolution of such curves. At lower heating rates, type (v) curves may tend to resemble type (iv) curves more closely, while at high heating rates both type (iv) and type (v) curves may resemble type (iii) curves and hence the detail of the complex decomposition is lost.

Type (vi) curves. These show a gain in mass as a result of reaction of the sample with the surrounding atmosphere. A typical example would be the oxidation of a metal sample.

Type (vii) curves. These are not often encountered. The product of an oxidation reaction decomposes again at higher temperatures (e.g. $2Ag + \frac{1}{2}O_2 \rightarrow Ag_2O \rightarrow 2Ag + \frac{1}{2}O_2$).

Resolution of stages of more complex TG curves can be improved by recording DTG curves (Fig. 3.8). Such curves may also be produced from digital TG data by numerical differentiation.

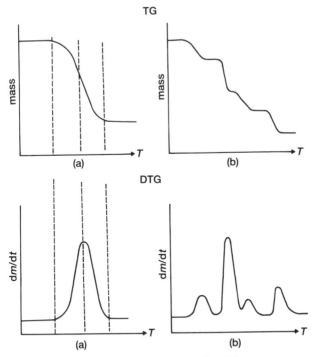

Figure 3.8 Comparison of TG and DTG curves.

3.9 Applications of TG

Application of TG is limited, to some extent, in that not all of the thermal events listed in chapter 2 are accompanied by changes in mass. For desorption, decomposition and oxidation processes, however, much valuable information can be obtained from TG alone. Much of the earlier work in TG was on the accurate definition of conditions for drying or ignition of analytical precipitates [7]. Examples of TG curves for $CuSO_4 . 5H_2O$ and for $CaSO_4 . 2H_2O$ are given in Figs 3.9 and 3.10. The mass losses define the stages, and the conditions of temperature (and surrounding atmosphere) necessary for preparation of the anhydrous compounds, or intermediate hydrates, can be established immediately. At higher temperatures, the sulphates will decompose further. Knowledge of the thermal stability range of materials provides information on problems such as the hazards of storing explosives, the shelf-life of drugs and the conditions for drying tobacco and other crops. By using an atmosphere of air or oxygen, the conditions under which oxidation of metals and degradation of polymers become catastrophic can be determined.

TG curves for more complex materials, such as minerals and polymers, are

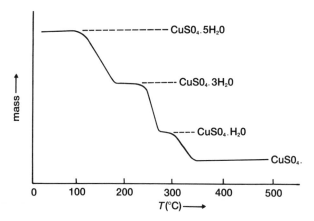

Figure 3.9 TG curve for $CuSO_4 \cdot 5H_2O$.

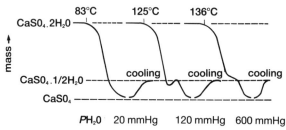

Figure 3.10 TG curve for $CaSO_4 \cdot 2H_2O$ at different water-vapour pressures. (Wiedemann, H. G. (1977) *J. Thermal Anal.*, **12**, 147, with the permission of Wiley–Heyden Ltd.)

not always immediately interpretable in terms of the exact reactions occurring. Such curves can, however, be used for 'fingerprint' purposes. The TG curve obtained on a given apparatus, under specified conditions, is compared with a bank of reference curves accumulated on the apparatus concerned. Some sets of reference curves have been published [30] but, as mentioned earlier, comparison is best when curves from the same instrument are available.

The reactions corresponding to the mass losses can best be determined, or confirmed, by simultaneous evolved gas analysis (EGA) (chapter 9). For example, in Fig. 3.9, the appearance of traces of SO_3, SO_2 and O_2 in the evolved gases would indicate the onset of sulphate decomposition. A complementary technique, such as hot-stage microscopy (HSM) (chapter 5) may provide information on the mechanism of dehydration or decomposition, by showing up the formation and growth of decomposition nuclei, or progress of the reactant/product interface inwards from crystal surfaces.

When the process occurring is clearly defined, e.g. the stoichiometric dehydration of a definite hydrate, the kinetics of the reaction can be determined from the TG curves (or from a series of isothermal curves of mass loss against time, obtained at different temperatures). Details of some of the kinetic analyses that have been suggested are given in chapter 13. Values of activation energies, obtained in this way, have been used to extrapolate to conditions of very slow reaction at low temperatures (in predicting shelf-lives of materials, resistance to weathering, and in estimating rates of natural processes, e.g. petroleum genesis over geological times) and to very fast reaction at high temperatures (behaviour of propellants and explosives).

The thermobalance may also be used to measure the vapour pressure of a sample, e.g. a metal, by determining the rate of mass loss through a calibrated orifice in a Knudsen cell.

It is worth noting, too, that the balance itself may be used, with little or no modification, for magnetic susceptibility measurements (chapter 11), accurate density measurements, determination of surface areas by adsorption, and particle-size analysis by sedimentation.

3.10 References

1. Czanderna, A. W. and Wolsky, S. P. (1980) *Microweighing in Vacuum and Controlled Environments*, Elsevier, Amsterdam.
2. Keattch, C. J. and Dollimore, D. (1975) *An Introduction to Thermogravimetry*, Heyden, London, 2nd edn.
3. Daniels, T. (1973) *Thermal Analysis*, Kogan Page, London.
4. Brown, M. E., Dollimore, D. and Galwey, A. K. (1980) *Reactions in the Solid State*, *Comprehensive Chemical Kinetics*, Vol. 22, (eds. C. H. Bamford and C. F. H. Tipper), Elsevier, Amsterdam.
5. Termeulen, J. P., Van Empel, F. J., Hardon, J. J., Massen, C. H. and Poulis, J. A. (1972) *Prog. Vacuum Microbalance Techniques*, Vol. 1, (eds. T. Gast and E. Robens), Heyden, London, p. 41.
6. Boersma, F. and Van Empel, F. J. (1975) *Prog. Vacuum Microbalance Techniques*, Vol. 3, (eds. C. Eyraud and M. Escoubes), Heyden, London, p. 9.
7. Henderson, D. E., DiTaranto, M. B., Tonkin, W. G., Ahlgren, D. J., Gatenby, D. A. and Shum, T. W. (1982) *Anal. Chem.*, **54**, 2067.
8. Okhotnikov, V. B. and Lyakhov, N. Z. (1984) *J. Solid State Chem.*, **53**, 161.
9. Offringa, J. C. A., de Kruif, C. G., van Ekeren, P. J. and Jacobs, M. G. H. (1983) *J. Chem. Thermodyn.*, **15**, 681.
10. Mulder, B. J. (1984) *J. Phys. E: Sci. Instrum.*, **17**, 119.
11. Braddick, H. J. J. (1963) *The Physics of Experimental Method*, Chapman and Hall, London, 2nd edn, p. 145.
12. Kishi, A., Takaoka, K. and Ichihasi, M. (1977) *Thermal Analysis*, Proc. 5th ICTA, (ed. H. Chihara), Heyden, London, p. 554.
13. Maesono, A., Ichihasi, M., Takaoka, K. and Kishi, A. (1980) *Thermal Analysis*, Proc. 6th ICTA, (ed. H. G. Wiedemann), Birkhauser, Basel, Vol. 1, p. 195.
14. Karmazin, E., Barhoumi, R., Satre, P. and Gaillard, F. (1985) *J. Thermal Anal.*, **30**, 43; (1984), **29**, 1269.

15. Karmazin, E., Barhoumi, R. and Satre, P. (1985) *Thermochim. Acta*, **85**, 291.
16. Cielo, P. (1985) *J. Thermal Anal.*, **30**, 33.
17. Steinheil, E. (1972) *Prog. Vacuum Microbalance Techniques*, Vol. 1, (eds. T. Gast and E. Robens), Heyden, London, p. 111.
18. Robens, E. (1985) *Vacuum*, **35**, 1.
19. Newkirk, A. E. (1971) *Thermochim. Acta*, **2**, 1.
20. Garn, P. D. and Alamalhoda, A. A. (1985) *Thermochim. Acta*, **92**, 833; (1982) *Thermal Analysis*, Proc. 7th ICTA, (ed. B. Miller), Wiley, New York, Vol. 1, p. 436.
21. Garn, P. D. and Kenessy, H. E. (1981) *J. Thermal Anal.*, **20**, 401.
22. Koppius, A. M., Poulis, J. A., Massen, C. H. and Jansen, P. J. A. (1972) *Prog. Vacuum Microbalance Techniques*, Vol. 1, (eds. T. Gast and E. Robens), Heyden, London, p. 181.
23. Schurman, J. W., Massen, C. H. and Poulis, J. A. (1972) *Prog. Vacuum Microbalance Techniques*, Vol. 1, (eds. T. Gast and E. Robens), Heyden, London, p. 189.
24. Cox, M. G. C., McEnaney, B. and Scott, V. D. (1973) *Prog. Vacuum Microbalance Techniques*, Vol. 2, (eds. S. C. Bevan, S. J. Gregg and N. D. Parkyns), Heyden, London, p. 27.
25. Oswald, H. R. and Wiedemann, H. G. (1977) *J. Thermal Anal.*, **12**, 147.
26. Boldyrev, V. V., Bulens, M. and Delmon, B. (1979) *The Control of the Reactivity of Solids*, Elsevier, Amsterdam.
27. Norem, S. D., O'Neill, M. J. and Gray, A. P. (1970) *Thermochim. Acta*, **1**, 29.
28. Elder, J. P. (1982) *Thermochim. Acta*, **52**, 235.
29. Duval, C. (1963) *Inorganic Thermogravimetric Analysis*, Elsevier, Amsterdam, 2nd edn.
30. Liptay, G. (ed.) (1971) *Atlas of Thermoanalytical Curves*, Vol. 1, (1973) Vol. 2, (1974) Vol. 3, (1975) Vol. 4, (1976) Vol. 5 and Cumulative Index, Heyden, London.
31. Blaine, R. L. and Fair, P. G. (1983) *Thermochim. Acta*, **67**, 233.
32. Gallagher, P. K. and Gyorgy, E. M. (1986) *Thermochim. Acta*, **109**, 193.
33. Charsley, E. L., Warne, S. St. J. and Warrington, S. B. (1987) *Thermochim. Acta*, **114**, 53.

Chapter 4

Differential thermal analysis (DTA) and differential scanning calorimetry (DSC)

4.1 Classical DTA [1, 2]

DTA is the simplest and most widely used thermal analysis technique. The difference in temperature, ΔT, between the sample and a reference material is recorded while both are subjected to the same heating programme. In 'classical' DTA instruments, represented schematically in Fig. 4.1, a single block with symmetrical cavities for the sample and reference is heated in the furnace. The block is chosen to act as a suitable heat-sink, and a sample holder of low thermal conductivity is included between the block and the sample to ensure an adequate differential temperature signal during a thermal event.

Should an endothermic thermal event (ΔH positive, such as melting) occur in the sample, the temperature of the sample, T_s, will lag behind the temperature of the reference, T_r, which follows the heating programme. If the output from the thermocouples, $\Delta T = T_s - T_r$, is recorded against T_r (or the furnace temperature, $T_f \simeq T_r$), the result will be similar to Fig. 4.1(c). If an exothermic process (ΔH negative, such as oxidation) occurs in the sample, the response will be in the opposite direction. Since the definition of ΔT as $T_s - T_r$ is rather arbitrary, each DTA curve should be marked with either the endo or exo direction. The negative peak, shown in Fig. 4.1(c), is called an **endotherm** and is characterized by its onset temperature. The temperature at which the recorder response is at its maximum distance from the baseline, T_{max}, is often reported but is very dependent upon the heating rate, ϕ, used in the temperature programme and factors such as sample size and thermocouple position. The area under the endotherm (or exotherm) is related to the value of the enthalpy change, ΔH, for the thermal event (for a more detailed interpretation, see section 4.7).

The reference material should have the following characteristics:
(i) It should undergo no thermal events over the operating temperature range.

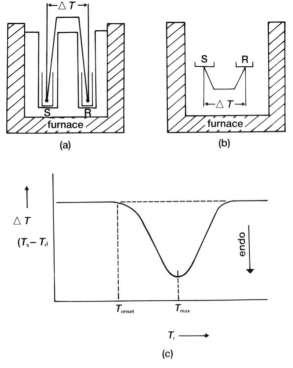

Figure 4.1 Differential thermal analysis (DTA). (a) Classical apparatus (S = sample; R = reference); (b) heat-flux; (c) typical DTA curve. (Note the DTA convention that endothermic responses are represented as negative, i.e. downward peaks.)

(ii) It should not react with the sample holder or thermocouple.
(iii) Both the thermal conductivity and the heat capacity of the reference should be similar to those of the sample.

Alumina, Al_2O_3, and carborundum, SiC, have been extensively used as reference substances for inorganic samples, while for organic compounds, especially polymers, use has been made of octyl phthalate and silicone oil.

Both solid samples and reference materials are usually used in powdered form. The particle size and the packing conditions influence results. A common technique for matching the thermal properties of the sample to those of the reference, is to use the reference material as a diluent for the sample. There must obviously be no reaction of the sample with the reference material.

The furnace system is usually purged with an inert gas and the possibilities of atmosphere control are similar to those discussed for TG (chapter 3).

4.2 Calorimetric DTA or heat-flux DSC [3]

In 'calorimetric' DTA, also known as Boersma DTA or differential dynamic calorimetry, the sample and reference, in similar holders, usually flat pans, are placed on individual thermally conducting bases. The thermocouple junctions are attached to these bases and are thus not directly in the sample or reference material. This configuration has the advantage that the output signal is less dependent upon the thermal properties of the sample (see above), but response is slower (Fig. 4.1(b)).

The temperature range of DTA depends on the materials used for the furnace windings and for the thermocouples.

4.3 Differential scanning calorimetry (DSC) [4, 5]

In power-compensated DSC, the sample and a reference material are maintained at the same temperature ($\Delta T = T_s - T_r = 0$) throughout the controlled temperature programme. Any **energy difference** in the independent supplies to the sample and the reference is then recorded against the programme temperature. The apparatus is shown schematically in Fig. 4.2(a), and an example of a resulting DSC curve in Fig. 4.2(b). The Perkin-Elmer DSC-2 is shown in Fig. 4.3 and a block diagram in Fig. 4.4. There are many similarities between DSC and DTA, including the superficial appearance of the thermal analysis curves obtained, but the principle of power-compensated DSC is distinctly different to that of calorimetric DTA, also known as heat-flux DSC (section 4.2).

Thermal events in the sample thus appear as deviations from the DSC baseline, in either an endothermic or exothermic direction, depending upon whether more or less energy has to be supplied to the sample relative to the reference material. Again, these directions should be clearly marked on the record, to avoid later confusion. In DSC, endothermic responses are usually represented as being positive, i.e. above the baseline, corresponding to an increased transfer of heat to the sample compared to the reference. Unfortunately this is exactly the opposite convention to that usually used in DTA, where endothermic responses are represented as negative temperature differences, below the baseline, as the sample temperature lags behind the temperature of the reference.

The operating temperature range of power-compensated DSC instruments is generally more restricted than that of DTA instruments. The Perkin-Elmer DSC-2, for example, has a maximum temperature of 726°C (999 K).

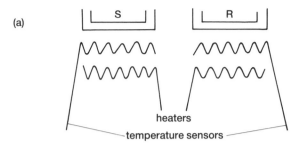

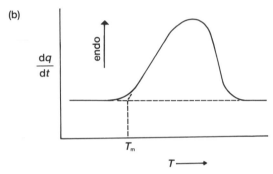

Figure 4.2 Differential scanning calorimetry (DSC). (a) Apparatus (S = sample; R = reference), (b) typical DSC curve. (Note the DSC convention, which is opposite to the DTA convention, that an endothermic response is represented by a positive, i.e. upward peak.)

Figure 4.3 Differential scanning calorimeter. (With the permission of Perkin-Elmer Corporation.)

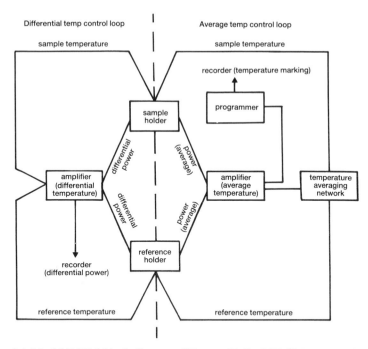

Figure 4.4 Model DSC-2 block diagram. (Watson, E. S., O'Neill, M. J., Justin, J. and Brenner, N. (1964) *Anal. Chem.*, **36**, 1233, with the permission of the American Chemical Society.)

Table 4.1 Recommended standard materials for calibration

(a) Temperature calibration

Material	Transition temperature (°C)	Material	Transition temperature (°C)
KNO₃	127.7	K₂SO₄	583
KClO₄	299.5	K₂CrO₄	665
Ag₂SO₄	412	BaCO₃	810
SiO₂ (quartz)	573	SrCO₃	925

(b) Enthalpy (and temperature) calibration

Material	M.pt(°C)	M.pt(K)	$\Delta H_{melting}$(cal g^{-1})	$\Delta H_{melting}$(J g^{-1})
Indium	156.4	429.6	6.80	28.5
Tin	231.9	505.1	14.40 ± 0.01	60.25 ± 0.04
Lead	327.4	600.6	5.45 ± 0.01	22.80 ± 0.04
Zinc	419.5	692.7	25.9 ± 0.1	108.4 ± 0.4
Aluminium	660.4	933.6	94.9 ± 0.3	397 ± 1

Taken from ref. 4.

4.4 Comparison of the principles of DTA and DSC [3, 6, 7]

A schematic diagram of a differential thermal apparatus is given in Fig. 4.5 together with definitions of the terms needed for comparison of classical DTA, heat-flux DSC and power-compensated DSC.

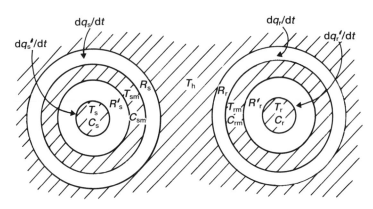

T_h = temperature of heat source

T_s	= actual sample temperature		T_r	= actual reference temperature
T_{sm}	= measured sample temperature		T_{rm}	= measured reference temperature
C_s	= heat capacity of sample + pan		C_r	= heat capacity of reference + pan
C_{sm}	= heat capacity of monitoring station		C_{rm}	= heat capacity of monitoring station

R'_s, R_s, R_r and R'_r are thermal resistances
The dq/dt terms are heat flows

Figure 4.5 Differential thermal apparatus. (From Mraw, S. C. (1982) *Rev. Sci. Inst.*, **53**, 228, with the permission of the American Institute of Physics.)

For an ideal instrument, the heat capacities and thermal resistances would be matched, i.e. $C_{rm} = C_{sm}$; $R_r = R_s = R$ and $R'_r = R'_s = R'$. Note that $R \neq R'$ and $C_s \neq C_r$. It is further assumed that $C_s > C_r$ and that heat flow is governed by Newton's law

$$dq/dt = (1/R)\Delta T$$

Heat flow to the sample side heats both (i) the sample monitoring station and (ii) the sample

$$dq_s/dt = C_{sm}(dT_{sm}/dt) + C_s(dT_s/dt)$$

also

$$dq'_s/dt = C_s(dT_s/dt)$$

so

$$dq_s/dt = C_{sm}(dT_{sm}/dt) + dq'_s/dt$$

Applying Newton's law

$$dq_s/dt = (1/R)(T_h - T_{sm})$$

and

$$dq'_s/dt = (1/R')(T_{sm} - T_s)$$

Similar expressions hold for the reference side.

Classical DTA

Thermocouples are in the sample and in the reference material so that $T_{sm} = T_s$ and $T_{rm} = T_r$, i.e. $R' = 0$.

The signal then is

$$\Delta T = R(dT/dt)(C_s - C_r)$$

and depends upon the difference in heat capacities, the heating rate and the thermal resistance, R. R is difficult to determine as it depends on both the instrument and the properties of the sample and the reference.

Power-compensated DSC

The power is varied to make $T_{sm} = T_{rm}$. Thus $T_h = T_{sm} = T_{rm}$ and $R = 0$, i.e. there is no thermal resistance. The signal is then $\Delta(dq/dt) = (dT/dt)(C_s - C_r)$.

Heat-flux DSC

The conditions are that

$$T_h \neq T_{sm} \neq T_s \text{ and } R \neq R' \neq 0.$$

The signal

$$\Delta T = T_{rm} - T_{sm} = R(dT/dt)(C_s - C_r)$$

which is of similar form to classical DTA, except that R depends only on the instrument and not on the characteristics of the sample.

R is a function of temperature and the relationship between R and T has to be determined by calibration of the instrument (section 4.6) at several different temperatures. Some manufacturers of heat-flux DSC instruments have incorporated electronic compensation for this temperature dependence and conversion of the instrument output to units of power, in the instrument control module.

4.5 Sample containers and sampling

At temperatures below 500°C (773 K) samples are usually contained in aluminium sample pans. One type of pan has lids which may be crimped (Fig. 4.6) into position, while for volatile samples, pans and press are available which enables a cold-welded seal, capable of withstanding 2–3 atm (220–300 kPa) pressure, to be made. It is worth noting that use of aluminium pans above 500°C will result in destruction of the DSC sample holder (several thousand pounds!). It is essential to ensure that the instrument is fitted with a protective circuit, limiting the maximum temperature which may be reached without operator intervention, and to inform the operators of the danger.

For temperatures above 500°C, or for samples which react with the Al pans, gold or graphite pans are available (but expensive!). Glass sample holders have been described in the literature but are not available commercially. The reference material in most DSC applications is simply an empty sample pan. The sample-holder assembly is purged by a gas which may be inert or reactive as desired. (The high thermal conductivity of helium makes it unsuitable for thermal measurements although it has advantages for evolved gas analysis.) Low temperature attachments are available for extending the operating range of the DSC to as low as −175°C.

Dense powders or discs, cut out of films with a cork borer, are ideal samples. Low density powders, flocks or fibres may be prepacked in a small piece of degreased aluminium foil to compress them. A small syringe is used for filling pans with liquid samples. When volatile products are to be formed on heating the sample, the pan lid should be pierced. This is also necessary when reaction with the purge gas is being investigated.

It is recommended that the total mass of the sample + pan + lid be recorded before and after a run so that further deductions on the processes occurring can be made from any change in mass.

4.6 Quantitative aspects of DTA and DSC curves

The features of interest in DTA or DSC curves are the deviations of the signal from the baseline, and the baseline is not always easy to establish. Initial displacements of the baseline itself from zero result from mismatching of the thermal properties of the sample and the reference material, and asymmetry in the construction of sample and reference holders. In severe cases this may cause

Figure 4.6 Crimping and welding presses. (With the permission of Perkin-Elmer Corporation and Mettler Instruments AG.)

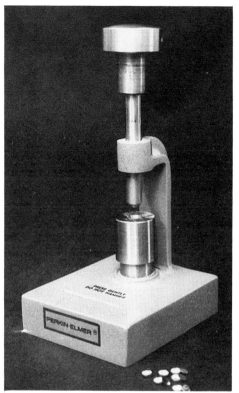
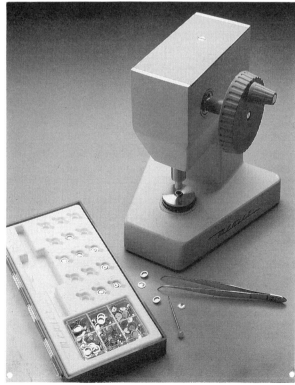

a sloping baseline, which may require electrical compensation. After a thermal event, the response will not return to the original baseline if the thermal properties of the high-temperature form of the sample are different from those of the low-temperature form. Many procedures for baseline construction have been suggested. Some of the simpler approximations are illustrated in Fig. 4.7.

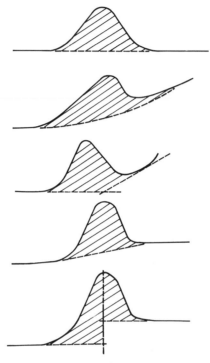

Figure 4.7 Simpler procedures for baseline extrapolation.

It should be immediately obvious whether a DSC or DTA feature is a peak or a discontinuity, provided that the baseline can be established with certainty. Whether a peak is in the exothermic or endothermic direction should be ascertained by comparison with a known melting (endothermic) peak. Although this seems obvious, reversal of polarity on the recorder switch or after temporary disconnection of the recorder, can cause a lot of confusion.

If there is any suggestion from initial runs that any of the features involve overlapping peaks, then further experiments should be done in which the sample mass and the heating rate are varied to see whether the individual peaks can be resolved.

Abrupt changes in slope or position of the baseline usually indicate **second-order transitions**. Examples of this type of transition are the glass transition in polymers (Fig. 4.8) and the Curie point transition in ferromagnetic materials (chapter 3). The enthalpy change, ΔH, for such transitions is zero, but there is a change in heat capacity.

Once a satisfactory baseline has been defined, the area of the endotherm or exotherm is determined by: (i) counting of squares; (ii) cutting out and weighing; (iii) planimetry; (iv) digitization and numerical integration; (v) mechanical or electronic integration. The measured area, A, is then assumed to be proportional to the enthalpy change, ΔH, for the thermal event represented:

$$\Delta H = A \times K/m$$

where m is the sample mass and K is the calibration factor. This factor has to be determined by relating a known enthalpy change to a measured peak area.

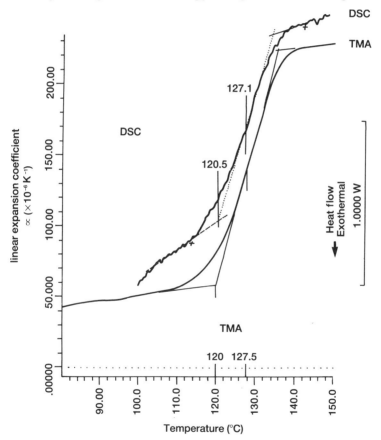

Figure 4.8 Glass transition of a printed circuit board (brominated bisphenol A resin). The DSC curve (proportional to the specific heat) is compared with the curve of the linear expansion coefficient (proportional to the first derivative of the TMA curve). (From Mettler Application No. 3402, with permission.)

Usually the melting of a pure metal, such as indium, is used for calibration. (The melting point of the standard is used, at the same time, to check the temperature calibration of the instrument.)

The calibration factor contains contributions from the geometry and thermal conductivity of the sample/reference system and is thus specific to a particular instrument under one set of operating conditions. For DTA or heat-flux DSC, the value of K is markedly dependent upon temperature, so that calibrations have to be carried out over the full operating temperature range of the instrument. Electronic compensation for the temperature dependence of the calibration factor is built into some heat-flux DSC instruments. A significant advantage of power-compensated DSC over DTA is that the proportionality constant, K, is virtually independent of temperature, and calibration is thus much easier.

Table 4.1 lists the transition temperatures and enthalpies of materials recommended as standards. A range of higher-temperature standards is likely to be available soon, through ICTA. Ideally, the thermal properties of the standard including its transition temperature, should be as close as possible to those of the sample under consideration.

The procedure for the determination of the enthalpy change for a process occurring in the sample, ΔH_s, is then to determine

$$K = \Delta H_{fusion,calibrant(c)} \times mass_c / A_c$$

and to use this value in

$$\Delta H_s = (KS_c R_s A_s)/(S_s R_c m_s)$$

where A is the area and subscripts c and s refer to calibrant and sample, respectively; S is the recorder chart speed and R the sensitivity range of the instrument/recorder. Note that the heating rate used in the temperature programme does not enter into the calculations, but, in precise work, should be the same for sample and calibrant.

4.7 Interpretation of DTA and DSC curves

At the heart of all thermal analysis experiments lies the problem of correlating the features recorded with the thermal events taking place in the sample. Some aspects of this correlation for TG results have been discussed in section 3.8. DSC or DTA provides different information and if simultaneous measurements (chapter 9) are not possible, the results of parallel experiments, using different techniques such as TG and DSC or DTA, under conditions (e.g. sample mass, heating rate, atmosphere) as closely matched as possible, are most valuable.

Once the main features of the DTA or DSC curve have been established, and baseline discontinuities have been examined, attention can be directed to the correlation of the endothermic or exothermic peaks with thermal events in the sample.

A useful procedure is to test whether the event being monitored is readily reversible on cooling and reheating, or not. Exothermic processes are not usually readily reversible, if at all, in contrast to melting and many solid–solid transitions.

The melting endotherms for pure substances are very sharp (i.e. they occur over a narrow temperature interval) and the melting point, T_m, is usually determined, as shown in Fig. 4.2(b), by extrapolation of the steeply rising, approximately linear, region of the endotherm back to the baseline. For impure substances the endotherms are broader and it is possible to estimate the impurity content (up to a maximum of about 3 mol%) from the detailed shape of the melting endotherm (chapter 14).

TG and EGD or EGA information is invaluable at this stage in distinguishing between irreversible or slowly-reversible phase transitions and decompositions. Quantitative mass losses and detailed EGA may, in addition, lead to determination of the stoichiometry of decomposition. Account must naturally be taken of the gaseous atmosphere surrounding the heated sample. Dehydration may be found to be reversible on cooling in a moist atmosphere before reheating, and carbonate decompositions are usually reversible in CO_2 atmospheres. Comparison of features observed, under otherwise similar conditions, in inert and oxidizing atmospheres is valuable. TG results may show increases in mass corresponding to reaction of the sample with the surrounding atmosphere.

It cannot be over emphasized that as many additional techniques available, such as hot-stage microscopy (chapter 5), conventional elemental analysis, X-ray diffraction (XRD) and the many types of spectroscopy, should be used to confirm suggested interpretations of TA features recorded. Table 4.2 summarizes the overall interpretation procedure, but this should not be followed blindly. There is no substitute for experience built up by running samples with well-documented behaviour on your own instrument, and also examining the almost legal-type of 'proof' of events provided by more conscientious authors (spurred on by more demanding editors).

4.8 Measurement of heat capacity

DSC can be used to measure heat capacities of materials. As heat capacity values are fundamental to thermodynamics, this is an important feature.

$$\Delta H = \int_{T_1}^{T_2} C_p \, dT \qquad \Delta S = \int_{T_1}^{T_2} (C_p/T) dT \qquad \Delta G = \Delta H - T\Delta S$$

where C_p is the heat capacity at constant pressure.

Table 4.2 Interpretation of DSC and DTA experiments

Feature	Peak				Discontinuity	
Direction	*Endothermic*		*Exothermic*			
Reversible	Yes	No	Yes	No	Yes	No
Broad	No	Yes	Usually	Usually	–	–
Mass loss (TG)	No	Yes	No	Yes (gain)	No	–
Gas evolved (EGD/EGA)	No	Yes	No	Yes	No	–
Possible interpretation	Solid transition or melting	Dehydration Decomposition (e.g. loss of ligand)	Polymer crystallization Some solid transitions	Decomposition (Oxidation)	Glass transition	Electrical disturbance
Further tests	Microscopy	XRD Spectroscopy	XRD	XRD	TMA DMA	Check recorder and shielding

In the absence of a sample, i.e. with empty pans in both holders of Fig. 4.2(a), the DSC baseline should be a horizontal line. A sloping baseline may be observed when the sample and reference holders have different emissivities, i.e. if the amount of energy lost by radiation does not vary in the same way with T_1 for both sample and reference. This could occur for a black sample and a shiny reference, so a sloping baseline is often an indication that the sample-holder assembly should be cleaned by heating in air to high temperatures, **with aluminium pans removed**.

When a sample is introduced, the DSC baseline is displaced in the endothermic direction and the displacement, h, is proportional to the total heat capacity, C_p, of the sample. (The total heat capacity = mass × specific heat capacity.) This is illustrated in Fig. 4.9(a)

$$h \propto C_p \qquad \text{or} \qquad h = B\phi C_p$$

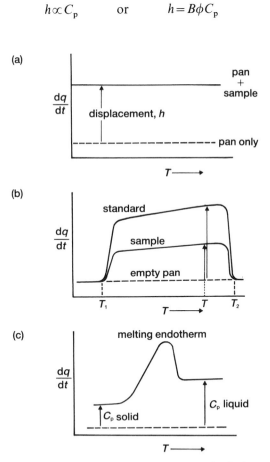

Figure 4.9 Determination of heat capacities using DSC. (a) Displacement of baseline, (b) scan of sample compared to standard, (c) shift in baseline after transition.

where ϕ is the heating rate and B the calibration factor. The value of B is determined using a standard substance, e.g. sapphire, scanned under similar conditions to the sample (Fig. 4.9(b)). Although the technique is simple in principle, Suzuki and Wunderlich [8] have shown that a great deal of care is necessary to get accurate and reproducible results.

Because of the proportionality of the DSC response to the heat capacity of the sample, a shift in baseline after a transition is common (Fig. 4.9(c)) and the baseline has to be estimated as shown in Fig. 4.7 or by more elaborate procedures [9].

4.9 Measurement of thermal conductivity

Thermal conductivity, λ, is an important physical property, particularly with the present emphasis on the efficient use of energy. Numerous methods for measuring thermal conductivity have been published and dedicated commercial instruments are available. Several papers have been published [10–13] in which heat-flux DSC instruments have been used to measure thermal conductivity. An advantage of using DSC is that specific heat capacities, c_p, can be measured in the same instrument (section 4.8) and hence, if density values, ρ, are available or can be measured, thermal diffusivities, D, can be calculated from the relationship, $D = \lambda/\rho c_p$.

Chiu and Fair [10] have described an attachment for the DuPont 990 DSC cell, (Fig. 4.10), which allows for measurement of thermal conductivity of a sample (in cylindrical form) without modification of the basic instrument. The temperature at the bottom of the sample (T_1) is measured *via* the output of the DSC, while the temperature at the top of the sample (T_2) is measured with a separate thermocouple in the contact rod. The heat input for the sample is also provided by the DSC. The cell is calibrated with a sample of known thermal conductivity. The output of the DSC operating in the normal mode is used as the recorder zero. With the sample in position, the DSC cell is brought to the desired measurement temperature, T_1 and when the output to the recorder is steady with time, the temperature difference across the sample, ΔT_s and the recorder displacement, h_s, are determined. The sample is then replaced by a calibrant, e.g. a standard glass, and the corresponding quantities, ΔT_c and h_c are determined. The thermal conductivity of the sample, λ_s, is then calculated from:

$$\lambda_s = \lambda_c \left(\frac{h_s R_s l_s d_c^2 \, \Delta T_c}{h_c R_c l_c d_s^2 \, \Delta T_s} \right)$$

where R is the recorder sensitivity, l the length and d the diameter of the sample (subscript s) and calibrant (subscript c) as specified.

A sample length of 10 to 25 mm is recommended and a precision of better

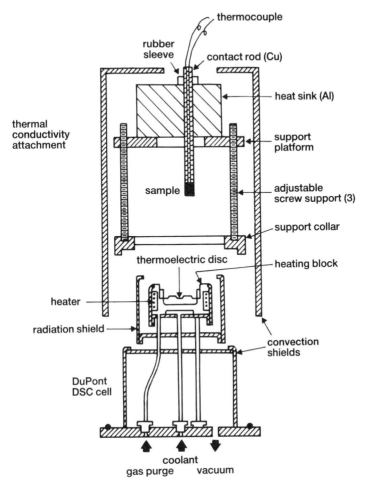

Figure 4.10 Schematic diagram of thermal conductivity cell [10]. (With the permission of *Thermochimica Acta*, Elsevier, Amsterdam.)

than 3% is claimed. Sircar and Wells [11] have used the above method in studying elastomer vulcanizates, and Hillstrom [12], a very similar system, for determining the thermal conductivity of explosives and propellants.

Hakvoort and van Reijen [13] using a DuPont 910 DSC, replaced the temperature sensor on the top of the sample by a disc of pure gallium or indium metal (Fig. 4.11(a)). The temperature at the top of the sample during the melting of the metal is then fixed at the melting point, T_m, of the metal. The y-output scale calibration factor C (in W mV^{-1}) for the instrument is first determined by melting a weighed sample of the metal placed directly on the

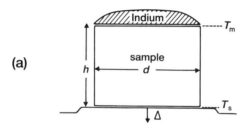

(a)

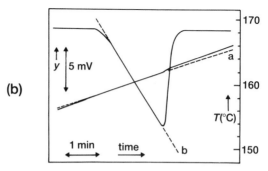

(b)

Figure 4.11 Measurement of thermal conductivity using DSC [13]. (a) sample geometry, (b) DSC plot, a: T_s against time→dT_s/dt, b: y against time→dy/dt; $(dT_s/dt)/(dy/dt) = dT_s/dy = \beta_p$. (With the permission of *Thermochimica Acta*, Elsevier, Amsterdam.)

sensor, without a sample pan, and measuring the DSC signal, y, with time. The heat of melting, Q_m, is given by

$$Q_m = \int Q' \, dt = C \int y \, dt,$$

so that

$$Q'_{sensor} = Cy$$

Two identical cylinders of sample (1–3 mm high, 2–4 mm diameter) are then placed on the sample and reference positions of the DSC (with silicone grease if necessary). A disc of the metal to be melted (10–40 mg and diameter similar to that of the sample) is placed on top of the cylinder on the sample side and the sensor temperature, T_s, and the DSC signal, y are recorded with time as the cell is heated at constant rate. During the melting of the metal, the top of the sample cylinder remains at constant temperature, T_m, while the lower side temperature, T_s, increases at constant rate. The DSC plot obtained is shown in Fig. 4.11(b).

Two quantities are determined: the slope of the plot of T_s against time, $a = dT_s/dt$, and the slope of the DSC curve, $b = dy/dt$. Combining these quantities, $a/b = dT_s/dy = \beta$. The thermal conductivity, λ, is then calculated from [4]

$$\lambda = hC/A\beta$$

where A is the cross-sectional area of the sample cylinder and h its height.

Boddington *et al.* [14] have carried out a more elaborate modification of a Stanton Redcroft DTA model 673 for measurement of the thermal diffusivity of pyrotechnic compositions.

4.10 Determination of phase diagrams

It is not always easy to detect accurately by eye the onset and completion of melting of binary or more complicated systems. The records obtainable from DSC and DTA runs, under carefully defined sample conditions, and slow heating and cooling rates so that equilibrium is approached, provide a more accurate way of establishing the phase diagram for the melting and solubility behaviour of the system under consideration.

In a binary system of immiscible solids and their completely miscible liquids, with the well-known phase diagram shown in Fig. 4.12 (e.g. the benzoic acid/naphthalene system [15, 16] or the triphenylmethane/*trans*-stilbene system [17]), the melting curves of the pure components A and B (with melting points $T_{m,A}$ and $T_{m,B}$) are readily obtained, using DSC or DTA, and should be sharply defined. The melting behaviour of mixtures of A and B will depend upon the histories of the mixtures and, in this discussion, it will be assumed that the mixtures have been prepared in completely molten form initially (without any volatilization or oxidation [18]).

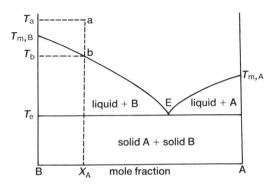

Figure 4.12 Phase diagram for a binary system of immiscible solids and their completely miscible liquids.

A DSC or DTA record of the slow cooling of such a molten mixture of composition x_A (mole fraction) initially at temperature T_a in Fig. 4.12, would show no deviation from the baseline until temperature T_b when solid B begins to crystallize in an exothermic process. This exotherm is not sharp but tails off as crystallization becomes complete (Fig. 4.13). The area under this exotherm will depend upon the amount of B present in the sample of the mixture. As the temperature of the mixture falls further, the eutectic temperature, T_e, is reached and solid A crystallizes in a sharp exotherm (Fig. 4.13). If the composition had been chosen to be that corresponding to point E, the DSC or DTA record would have shown only a single sharp exotherm as the eutectic composition solidified. At low concentrations of A, and similarly at low concentrations of B, the formation of the eutectic may be difficult to detect and hence to distinguish from the formation of solid solutions [18].

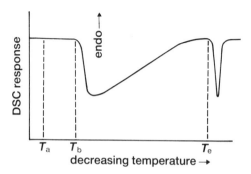

Figure 4.13 DSC or DTA record of the slow cooling of a molten mixture of composition x_A initially at temperature T_a in Fig. 4.12.

If the mixture, whose cooling curve is shown in Fig. 4.13, were to be reheated slowly, the DSC or DTA record should ideally be the endothermic mirror image of that shown. The use of the heating rather than the cooling curve avoids problems of supercooling.

Ideally then, the DSC or DTA traces should be readily relatable to the phase diagram, as illustrated schematically in Fig. 4.14. The area under the peak for the eutectic melting is a simple function of concentration (Fig. 4.15) and such a curve may be used to determine the composition of an unknown mixture, e.g. an alloy [19].

Pope and Judd [18] give an example of a more complicated phase diagram, with incongruently and congruently melting compounds and solid solution, eutectic and liquidus reactions, while Eysel [20] discusses the even more complex Na_2SO_4/K_2SO_4 system. A variety of systems, including ternary systems and phase studies under high pressure, is discussed by Gutt and Majumdar [21], and the theoretical background is covered by Sestak [22]. The

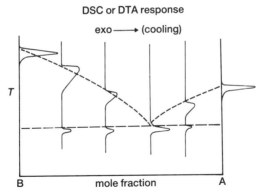

Figure 4.14 Schematic comparison of DSC or DTA traces with the phase diagram.

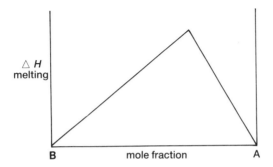

Figure 4.15 Relationship between enthalpy of melting and composition.

phase diagrams for the ternary Ge–Sb–Bi system [23] and the complicated Ag_2Te–Ag_4SSe system, with 17 phase regions [24], serve as illustrations of how successful such studies can be.

Combination of DTA with thermomicroscopy (chapter 5 and ref. 25) can provide additional information over a more limited temperature range.

4.11 General applications of DTA and DSC

The results obtained using DTA and DSC are qualitatively so similar that their applications will not be treated separately. It should be noted that DTA can be used to higher temperatures than DSC (max 725°C) but that more reliable quantitative information is obtained from DSC.

Once again, DTA and DSC curves may be used solely for 'fingerprint' comparison with sets of reference curves. It is usually possible, however, to extract a great deal more information from the curves, such as the temperatures and enthalpy changes for the thermal events occurring. As an example, the DSC curve for $CuSO_4 . 5H_2O$ is shown in Fig. 4.16 for comparison with the TG curve given in Fig. 3.9. Note the increase in ΔH of dehydration per mole of H_2O removed.

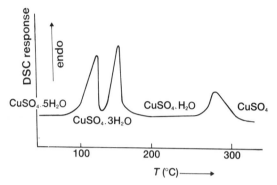

Figure 4.16 DSC curve for $CuSO_4 . 5H_2O$.

As mentioned earlier, the shape of the melting endotherm can be used to estimate the purity of the sample. The procedure is discussed in detail in chapter 14.

Detection of solid–solid phase transitions, and the measurement of ΔH for these transitions, is readily done by DSC or DTA. A low-temperature attachment permits the range of samples which can be examined to be extended, as shown by the DSC curve for carbon tetrachloride in Fig. 4.17. Satisfactory operation as low as $-175°C$ has been achieved. Condensation of atmospheric

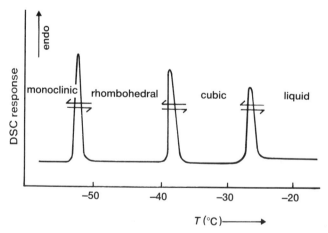

Figure 4.17 DSC curve for CCl_4.

moisture can cause problems and has to be reduced. Redfern [26] has reviewed some of the cooling systems available and has discussed some low-temperature applications.

Application of thermal analysis to the study of polymers has been most rewarding. A DSC curve for a typical organic polymer is shown in Fig. 4.18.

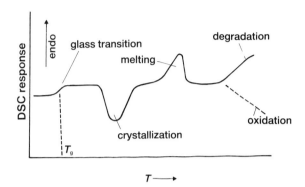

Figure 4.18 DSC curve for a typical organic polymer.

Most solid polymers are formed by rapid cooling to low temperatures (quenching) from the melt and are thus initially in the glassy state. The transition from a glass to a rubber, the **glass transition**, is an example of a second-order phase transition. Such transitions are accompanied by a change in heat capacity, but no change in enthalpy ($\Delta H = 0$). The transition thus appears on the DSC curve as a discontinuity in the baseline at the glass-transition temperature, T_g. As the temperature is slowly increased the polymer may recrystallize giving the exotherm shown, before melting occurs. At higher temperatures the polymer may decompose (degrade) or oxidize depending upon the surrounding atmosphere.

The degradation or oxidation of polymers can be studied using DSC in the isothermal mode. Figure 4.19 shows the effect of stabilizers on the oxidation of

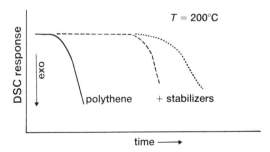

Figure 4.19 Effect of stabilizers on the oxidation of polythene, using isothermal DSC. (Cassel, B. (1975) *Ind. Res.*, Aug., 53–56, with permission.)

polyethylene at 200°C. The crystallization of polymers can also be studied using isothermal DSC. Several approaches are possible. The molten polymer may be cooled rapidly to a temperature in the crystallization range, and the crystallization from the liquid observed, or the temperature may be raised from ambient so that crystallization from the rubber-like state is observed.

Re-use of plastic waste is obviously most desirable, but is hampered by the problems of identification, sorting and collection. DSC may aid in identification of the constituents of the scrap, as illustrated in Fig. 4.20.

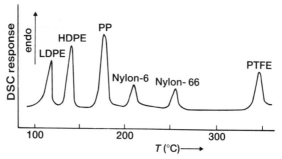

Figure 4.20 DSC of a sample of plastic waste (LDPE = low density polyethylene; HDPE = high density polyethylene; PP = polypropylene; PTFE = poly(tetrafluoro-ethylene), 'Teflon').

An example is given in Fig. 4.21 of the use of DSC in testing for completeness of curing of epoxy resins. The second scan shows no residual exotherm but indicates the glass-transition temperature for the cured resin.

The behaviour of liquid crystals is of practical importance because of the optical phenomena which occur at the transition points, and their proposed role in biological processes. Instead of going directly from solid to liquid at a sharp melting point, liquid crystals form several mesophases between solid and liquid states. The compounds which display this behaviour generally have large

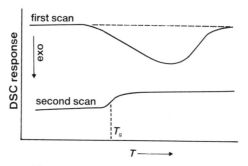

Figure 4.21 Curing of an epoxy resin.

asymmetric molecules with well-separated polar and non-polar regions. In the **smectic** mesophase molecules are aligned and tend to form similarly aligned layers; while in the **cholesteric** mesophase the orientation of the molecular axes shifts in a regular way from layer to layer, and in the **nematic** mesophase molecules are aligned, but layers are not formed (Fig. 4.22). Changes from one mesophase to another can be detected using DSC, and the enthalpies of transition evaluated, as shown for cholesteryl myristate in Fig. 4.23.

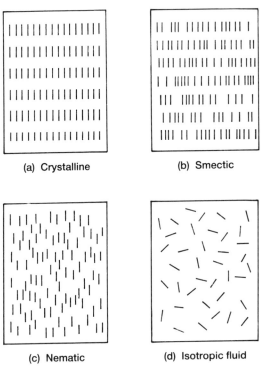

(a) Crystalline   (b) Smectic

(c) Nematic   (d) Isotropic fluid

Figure 4.22 Liquid crystal phases. (Based on Moore, Walter J. (1972) *Physical Chemistry*, 4th edn., with the permission of Prentice-Hall, Inc., Englewood Cliffs, New Jersey.)

DSC can also be used to study the liquid→vapour and solid→vapour transitions [27, 28] and to measure the enthalpies of vaporization and of sublimation. In open sample pans these changes are spread out over too wide a temperature interval for an accurate baseline to be determined. The rate of escape of vapour is reduced by partial sealing of the pan. This sharpens up the trace as shown in Fig. 4.24.

Structural changes in fats and waxes also show up clearly, making the technique (with low temperature attachments) ideal for food chemistry (Fig. 4.25).

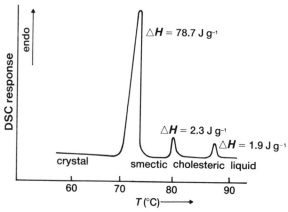

Figure 4.23 DSC curve of a liquid crystal (chloresteryl myristate) (Brennan, W. P. and Gray, A. P. (1974) *Perkin-Elmer Thermal Analysis Application Study*, No. 13, April, with permission.)

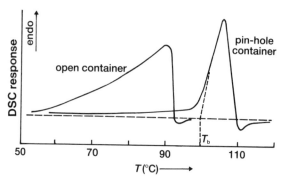

Figure 4.24 DSC curves for water [27, 28]. (From *Perkin-Elmer Thermal Analysis Newsletter* No. 7, 1967, with permission.)

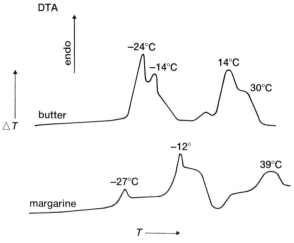

Figure 4.25 Butter and margarine – you can tell the difference!

4.12 References

1. Mackenzie, R. C. (ed.) (1969) *Differential Thermal Analysis*, Vol. 1 and 2, Academic Press, London.
2. Pope, M. I. and Judd, M. D. (1977) *Differential Thermal Analysis*, Heyden, London.
3. Sestak, J., Satava, V. and Wendlandt, W. W. (1973) *Thermochim. Acta*, **7**, 372.
4. McNaughton, J. L. and Mortimer, C. T. (1975) *Differential Scanning Calorimetry*, Perkin-Elmer Order No. L-604 (reprinted from IRS, Phys. Chem. Ser. 2, Vol. 10, Butterworths).
5. Fyans, R. L., Brennan, W. P. and Earnest, C. M. (1985) *Thermochim. Acta*, **92**, 385.
6. Mraw, S. C. (1982) *Rev. Sci. Inst.*, **53**, 228.
7. Saito, Y., Saito, K. and Atake, T. (1986) *Thermochim. Acta*, **99**, 299.
8. Suzuki, H. and Wunderlich, B. (1984) *J. Thermal Anal.*, **29**, 1369.
9. Brennan, W. P., Miller, B. and Whitwell, J. C. (1969) *Ind. Eng. Fundam.*, **9**, 314.
10. Chiu, J. and Fair, P. G. (1979) *Thermochim. Acta*, **34**, 267.
11. Sircar, A. K. and Wells, J. L. (1982) *Rubber Chem. Technol.*, **55**, 191.
12. Hillstrom, W. W. (1979) *Thermal Conductivity*, **16**, 483.
13. Hakvoort, G. and van Reijen, L. L. (1985) *Thermochim. Acta*, **93**, 371; (1985) **85**, 319.
14. Boddington, T., Laye, P. G. and Tipping, J. (1983) *Comb. Flame*, **50**, 139.
15. Visser, M. J. and Wallace, W. H. (1966) *DuPont Thermogram*, **3** (2), 9.
16. Daniels, T. (1973) *Thermal Analysis*, Kogan Page, London, p. 119.
17. McNaughton, J. L. and Mortimer, C. T. (1975) *Differential Scanning Calorimetry*, Perkin-Elmer Order No. L-604 (reprinted from IRS, Phys. Chem. Ser. 2, Vol. 10, Butterworths). p. 28.
18. Pope, M. I. and Judd, M. D. (1980) *Differential Thermal Analysis*, Heyden, London, p. 53.
19. Fyans, R. L. (1970) *Perkin-Elmer Instrument News*, **21** (1), 1.
20. Eysel, W. (1971) *Thermal Analysis*, Proc. 3rd ICTA, Vol. 2, 179.
21. Gutt, W. and Majumdar, A. J. (1972) *Differential Thermal Analysis*, (ed. R. C. Mackenzie), Academic, London, Vol. 2, p. 79.
22. Sestak, J. (1984) *Thermophysical Properties of Solids, Comprehensive Analytical Chemistry*, Vol. XIID, (ed. G. Svehla) Elsevier, Amsterdam, p. 109.
23. Surinach, S., Baro, M. D. and Tejerina, F. (1980) *Thermal Analysis*, Proc. 6th ICTA, Vol. 1, p. 155.
24. Boncheva-Mladenova, Z. and Vassilev, V. (1980) *Thermal Analysis*, Proc. 6th ICTA, Vol. 2, p. 99.
25. Perron, W., Bayer, G. and Wiedemann, H. G. (1980) *Thermal Analysis*, Proc. 6th ICTA, Vol. 1, p. 279.
26. Redfern, J. P. (1972) *Differential Thermal Analysis*, (ed. R. C. Mackenzie), Academic Press, London, Vol. 2, p. 119.
27. Daniels, T. (1973) *Thermal Analysis*, Kogan Page, London, p. 132.
28. McNaughton, J. L. and Mortimer, C. T. (1975) *Differential Scanning Calorimetry*, Perkin-Elmer Order No. L-604 (reprinted from IRS, Phys. Chem. Ser. 2, Vol. 10, Butterworths), p. 19.

Chapter 5

Thermoptometry

5.1 Introduction

Thermoptometry is a group of techniques in which optical properties of a sample are measured as a function of temperature [1]. Under this heading, the major technique is thermomicroscopy, i.e. direct observation of the sample. Other techniques involve measurement of total light reflected or transmitted (thermophotometry), light of specific wavelength(s) (thermospectrometry), refractive index (thermorefractometry), or emitted light (thermoluminescence).

5.2 Thermomicroscopy

It is always interesting that the most obvious property of a material, i.e. its outward appearance, is so often not examined or monitored during the heating process and yet can provide so much reliable first-hand evidence of the processes occurring [2]. Other thermal analysis techniques give results which have to be associated by inference with thermal events and some processes such as sintering, decrepitation and creeping and foaming of melts are only really detectable by direct observation. Controlled temperature stages for microscopes are readily available and can cover the temperature range from $-180°C$ up to 3000°C, allowing for study of a wide range of problems, especially phase studies (chapter 4). Care has to be taken that suitable heat-resistant objective lenses of adequate focal length are available. Standard photographic equipment or video cameras can be used for recording observations. Allowance may have to be made for heating by the illuminating source, or heat filters may be necessary.

If visual examination can be combined (chapter 9) with a quantitative measurement such as DSC or DTA, even more meaningful results can be obtained. The Mettler system [3, 4] (Fig. 5.1) has a heat-flux DSC sensor built from thin-film thermopiles deposited on a special glass disc. The sample is placed in a sapphire crucible and a similar but empty crucible is used as a reference. Sample illumination is by transmitted (normal or polarized) light.

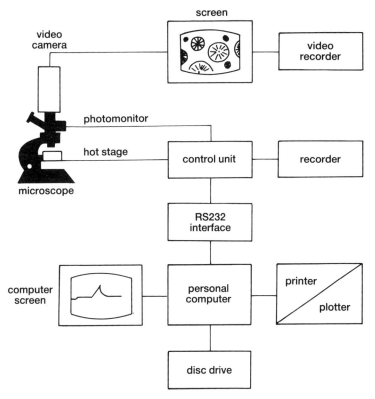

Figure 5.1 Instrumental set-up for simultaneous hot stage microscopy [3, 4]. (With the permission of Mettler Instrumente AG.)

The temperature range is from room temperature to 300°C. Wiedemann and Rossler [5] have reported on the use of this system for a detailed study of the dehydration of gypsum.

5.3 Thermophotometry

In thermophotometry provision is made for measurement of the intensity of the light reflected or transmitted by the sample. If samples are viewed in transmitted light using crossed polars, the only light transmitted is that arising from rotation of the plane of polarization of the light, caused by changes in the crystalline structure of the sample. Melting or the formation of an isotropic structure results in complete extinction of the light. The information provided is thus complementary to DTA or DSC measurements. Arneri and Sauer [6] have shown how such a technique can be used in the study of polymers, where overlapping transitions, such as melting and degradation or crystallization and melting, can lead to ambiguities of interpretation.

In the Stanton Redcroft system [7], the sample is contained between two microscope cover glasses which rest on a heating block, within an atmosphere-controlled chamber. The block contains a sapphire window for light transmission, and a platinum resistance thermometer for measurement of the temperature and control of the heating or cooling programme. Provision for cooling with liquid nitrogen gives an overall temperature range of $-180°C$ to $600°C$. The intensity of the transmitted light is measured using a silicon photodetector. An example of the results obtained on heating KNO_3 is shown in Fig. 5.2. The phase transition from orthorhombic, phase II, to trigonal, phase I at $128°C$ and melting at $335°C$ show up clearly. The system has also been used [8] for determining the ignition temperatures of pyrotechnic compositions. An earlier version of the hot-stage microscope [9] had a temperature range up to $1000°C$, but the sample had to be viewed in reflected light.

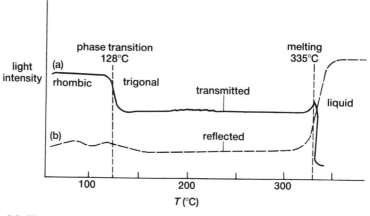

Figure 5.2 Thermophotometry curves for potassium nitrate [12]. (a) In transmitted light, (b) in reflected light. (With the permission of *Thermochimica Acta*, Elsevier, Amsterdam.)

Sommer and Jochens [10] have described a micro-thermal analysis system based on the use of a thermocouple as both specimen holder and heating source. The system is easily and cheaply constructed and is particularly useful in phase studies of non-metallic samples [11].

Haines and Skinner [12] have modified a DSC to carry out simultaneous measurements of the intensity of light reflected from the sample. Changes in the surface of the sample may not be accompanied by measurable enthalpy changes, but may show up clearly through changes in reflectance. A binocular microscope was used so that one eyepiece could be used for photographing the actual appearance of the sample, while the photodetector was used on the other eyepiece. The sample was in either an open DSC pan, or else a mica window was crimped into the pan. Changes in the intensity of the reflected light are cumulative so that the curves recorded are related to DSC curves as integral to derivative (compare TG and DSC). The phase transition at 128°C for KNO_3 did not show up as clearly as in transmitted light (Fig. 5.2). The transitions in liquid crystals were readily detectable. Dehydrations and decompositions were also accompanied by significant changes in reflected light intensity. The system also proved useful in the study of the browning and ignition of paper samples [12].

5.4 Thermoluminescence

The technique involving measurement of light actually being emitted by the sample is known as thermoluminescence, TL [13] and has been used, in particular, by Wendlandt in the study of coordination compounds [14, 15]. TL arises from the annealing of defects in the solid sample as the temperature is raised. The defects are present in the sample on account of its previous thermal history, or through irradiation of the sample. Light emission from polymers [14] may result from the formation of hydroperoxide radicals, while vigorous redox reactions between oxidizing groups and reducing ligands in coordination compounds may also be accompanied by light emission [14, 15].

Nuzzio [16] has described the modification of a DuPont 990 Thermal Analyzer for TL measurements including arrangements for computer data capture. The light-measuring system included filters to absorb heat and to minimize the black-body radiation encountered at temperatures above $\sim 400°C$. A reference light source was used for calibration. TL measurements could be made under isothermal or increasing temperature conditions. Arrangements for measurements under reduced pressures are also described.

Manche and Carroll [17] modified a Perkin-Elmer DSC-1B for simultaneous TL and DSC, using a matched pair of optical fibres mounted directly above each calorimeter cup. The system was tested using mixtures of LiF and KNO_3. LiF is a well-known and extensively-studied thermoluminescent material and KNO_3 was used as the temperature and calorimetric standard and does not normally exhibit thermoluminescence. Ground mixtures were pressed into discs and the discs were irradiated with X-rays. The sample discs, with a preheated disc of LiF as reference, were heated in the modified DSC. The results obtained are shown in Fig. 5.3. Areas under the glow curves and the DSC endotherms were measured and were shown to be linear functions of the mass fraction of the individual salts.

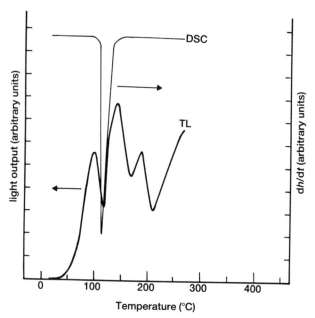

Figure 5.3 Thermoluminescence (TL) and DSC curves for mixtures of LiF and KNO$_3$ [17]. (With the permission of the American Chemical Society.)

5.5 Electron microscopy

Many of the kinetic studies of mechanisms of decomposition of solids (chapter 13) invoke processes of formation and growth of product nuclei. These speculations can often be confirmed or refuted using thermomicroscopy.

Use of scanning electron microscopy, rather than optical microscopy, has the advantages of greater magnification and much increased depth of field. There are, however, additional problems in the necessity for operating under vacuum and the effect of the electron beam on the sample. Many attempts have been made to study reactions directly in the electron microscope, but none of these so far has had the temperature control required for thermal analysis.

5.6 References

1. Mackenzie, R. C. (1978) *J. Thermal Anal.*, **13**, 387.
2. McCrone, W. C. (1976) *Proc. ESTA 1*, (ed. D. Dollimore) Heyden, London, p. 63.
3. Wiedemann, H. G. and Bayer, G. (1985) Proc. 8th ICTA, *Thermochim. Acta*, **92**, 399.

4. Perron, W., Bayer, G. and Wiedemann, H. G. (1980) *Thermal Analysis*, Proc. 6th ICTA, (ed. H. G. Wiedemann) Birkhauser Verlag, Basel, Vol. 1, p. 279.
5. Wiedemann, H. G. and Rossler, M. (1985) *Thermochim. Acta*, **95,** 145.
6. Arneri, G. and Sauer, J. A. (1976) *Thermochim. Acta*, **15,** 29.
7. Charsley, E. L., Kamp, A. C. F. and Rumsey, J. A. (1985) Proc. 8th ICTA, *Thermochim. Acta*, **92,** 285.
8. Charsley, E. L. and Kamp, A. C. F. (1971) *Thermal Analysis*, Proc. 3rd ICTA, (ed. H. G. Wiedemann) Birkhauser, Basel, Vol. 1, p. 499.
9. Charsley, E. L. and Ottaway, M. R. (1976) Proc. ESTA 1, (ed. D. Dollimore) Heyden, London, p. 444.
10. Sommer, G. and Jochens, P. R. (1971) *Minerals Sci. Eng.*, **3,** 3.
11. Glasser, L. and Miller, R. P. (1965) *J. Chem. Ed.*, **42,** 91.
12. Haines, P. J. and Skinner, G. A. (1982) *Thermochim. Acta*, **59,** 343.
13. Ball, M. C. and Marsh, C. M. (1985) *Thermochim. Acta*, **91,** 15.
14. Wendlandt, W. W. (1980) *The State-of-the-Art of Thermal Analysis*, NBS Spec. Tech. Publ. 580, (eds. O. Menis, H. L. Rook and P. D. Garn) p. 219.
15. Wendlandt, W. W. (1980) *Thermochim. Acta*, **35,** 247; (1980) **39,** 313.
16. Nuzzio, D. B. (1982) *Thermochim. Acta*, **52,** 245.
17. Manche, E. P. and Carroll, B. (1982) *Anal. Chem.*, **54,** 1236.

Chapter 6

Thermodilatometry (TD)

6.1 Basic principles

Measurement of the expansion of solids and liquids caused by various imposed conditions, such as absorption or chemical reaction or other time-related processes, at fixed temperatures is known as **dilatometry**. When special emphasis is put on recording such dimensional changes as a function of temperature, during a controlled temperature programme, the technique is labelled **thermodilatometry** [1] and is often included under the general heading of thermomechanical analysis (TMA) (chapter 7). For solids, we can distinguish between linear and volume expansion, and instruments for measuring the length of a solid sample as a function of temperature are the most common.

Most solids expand on heating. Exceptions include vitreous silica and ZnS at low temperatures. The change in a linear dimension, L, with T is given by

$$L_2 = L_1 \left(1 + \int_{T_1}^{T_2} \alpha \, dT \right)$$

where L_1 is the length at T_1, and L_2 the length at T_2, and α is the coefficient of linear expansion. Over a small temperature interval, ΔT, α is approximately constant for many materials, so the dilation

$$L_2 - L_1 = \Delta L = L_1 \, \alpha \, \Delta T$$

or

$$\Delta L / L_1 = \alpha \, \Delta T$$

The value of α is related to the structure and type of bonding in the solid. In general, strong bonding results in low values of α, so the order of expansion coefficients is usually: covalent and ionic materials < metals < molecularly-bonded materials. In crystalline substances, α may vary with direction (anisotropic behaviour).

Measurements of thermal expansion may be classed as either absolute or difference measurements, as illustrated schematically in Fig. 6.1. The pointer and scale are outside the furnace and the movements of the sample and of the

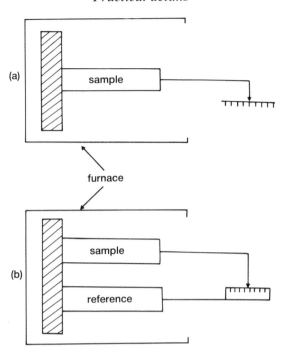

Figure 6.1 Thermodilatometry. (a) Absolute measurement, (b) difference measurement.

reference are transmitted by identical push-rods. Measurements may be made in either the horizontal or vertical position. In horizontal instruments, arrangement has to be made for suitable spring-loading, so that any contraction can be detected. In vertical instruments, the weight of the push-rod and measuring system has to be counterbalanced to avoid compression or deformation of the sample.

6.2 Practical details

Measurement of sample movement obviously has to be more sophisticated than the pointer and scale used in the schematic diagrams. The transducer (Fig. 1.1) is usually a linear variable differential transformer (LVDT), the principle of which is illustrated in Fig. 6.2. Movement of the core within the coils produces an output with sign dependent upon the direction of the movement, and magnitude giving the movement of the sample. Amplification of the movement is claimed to be as much as $20\,000 \times$. Calibration of the LVDT is carried out, initially, using a micrometer screw gauge incorporated in the measuring system.

Safety devices, to prevent damage caused by melting of the sample in the

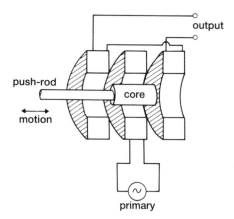

Figure 6.2 Principle of a linear variable differential transformer (LVDT).

apparatus, are often included. These are activated by rapid sample contractions and power to the furnace is cut off immediately.

Expansion of a standard material can be used as a temperature indicator. Isothermal measurements may be made to study creep or recrystallization processes in the sample (see applications). Changes in volume of liquid samples, contained in a suitable cylinder, can be converted into equivalent linear movement by means of a close-fitting piston attached to a push-rod. A bleed valve on the cylinder ensures that no air is trapped in the liquid.

Linseis [2] has described a laser dilatometer based on the principle of a Michelson interferometer. A schematic diagram is given in Fig. 6.3. The laser beam is split, after a 90° reflection, into two separate beams. The two beams are directed at reflectors attached to the ends of the sample and reference materials. After total reflection, the beams are recombined and the interference patterns are examined using photodiode detectors and converted to measurements of expansion or shrinkage. These dilation measurements are accurate to one-quarter of the wavelength of the laser and involve no mechanical connections

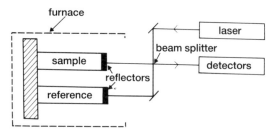

Figure 6.3 Laser dilatometer [2]. (With the permission of John Wiley & Sons, Chichester.)

with the sample, so very rapid changes of dimension can be examined. The temperature range over which such an optical system can be used is also much greater than for a mechanical system and it is suggested [2] that it will prove useful in examining the expansion coefficients of superconducting materials at liquid helium temperatures.

Karmazsin *et al.* [3] have suggested use of an optoelectronic transducer in place of an LVDT. This has particular advantages for simultaneous thermodilatometric, thermoconductimetric and thermomagnetometric measurements as LVDTs are sensitive to electrical perturbations. A stabilized light source (Fig. 6.4) illuminates two photosensitive cells through two rectangular holes of

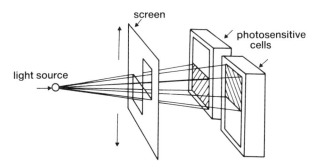

Figure 6.4 Optoelectronic transducer for thermodilatometry [3]. (With the permission of John Wiley & Sons, Chichester.)

a screen, which is attached to the push-rods of the dilatometer. Movement of the screen alters the relative amount of light received by the two cells. Dimensional variations down to 10^{-2} μm could be measured and speed of response was about ten times that of an LVDT. Examples of simultaneous measurements are given [3].

6.3 Interpretation of results

Values of the coefficient of expansion, α, may be determined from the slope of a curve of L against T, since

$$\Delta L/\Delta T = dL/dT = L_1 \alpha$$

Alternatively, the time derivative of the dilation may be recorded, using electronic differentiation, and

$$(dL/dt)(dt/dT) = L_1 \alpha$$

so

$$dL/dt = L_1 \alpha\phi \quad \text{where} \quad \phi = dT/dt = \text{heating rate.}$$

The engineering coefficient of expansion, α', is defined by

$$\alpha' = (L_T - L_{20})/(L_{20}(T-20))$$

where L_T and L_{20} are the lengths of the sample at $T°C$ and $20°C$, respectively.

Usually it is the discontinuities in the curve of L against T, rather than actual values of α, which are of main interest in thermodilatometry, as these indicate occurrence of thermal events in the sample. These discontinuities are discussed in the applications section below.

6.4 Applications of thermodilatometry

As mentioned earlier, thermodilatometry (TD) and thermomechanical analysis (TMA) (chapter 7) are closely related and they have been applied in similar fields (see also chapter 7). A very practical application of TD is in the selection of suitable materials for use as brake-linings [4]. Obviously a low coefficient of expansion, coupled with other qualities such as wear-resistance and a suitable coefficient of friction, is desirable (Fig. 6.5).

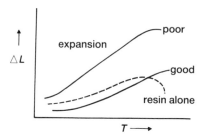

Figure 6.5 Thermal expansion of brake-lining materials [4]. (With the permission of International Scientific Communications, Inc.)

Wendlandt [5] has used TD to study the change in structure from octahedral to tetrahedral in the cobalt(II) pyridine coordination compound, $Co(py)_2Cl_2$ (Fig. 6.6).

Another application of TD has been that of Gallagher [6] to the study of the kinetics of sintering of powdered Chromindur alloys (Fe–Cr–Co), used in making high energy magnets. Arrhenius-type plots (chapter 13) of $\ln(L/L_0)$ against $1/T$ showed two regions with different activation energies. The values of the activation energies corresponded well with the diffusion energies in the low-

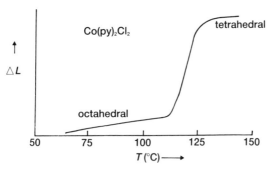

Figure 6.6 Thermodilatometry curve for Co(py)$_2$Cl$_2$ [5]. (With the permission of *Analytica Chimica Acta*, Elsevier, Amsterdam.)

temperature face-centred cubic and the high-temperature ($>1250°C$) body-centred cubic phases.

TD has also been used [7, 8] to study the sintering behaviour of kaolins and kaolinitic clays. On heating, kaolin loses structural water at about 550°C and forms the metakaolin structure. Above 960°C this converts to a spinel structure and then above 1100°C to mullite. The mullite formed sinters at higher temperatures. These reactions are accompanied by the shrinkages shown in Fig. 6.7. The TD curves of kaolins and kaolinitic clays may be used for their classification [7].

Some other recent applications of TD have been to study the effect of additives on the shrinkage of cellular concretes [9]; the defects in non-stoichiometric oxides [10]; the firing of ceramics and the properties of alloys [11].

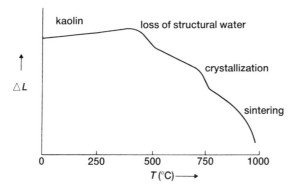

Figure 6.7 Thermodilatometry (TD) curve on kaolin [7, 8]. (With the permission of Fisher Scientific Co.)

6.5 References

1. Daniels, T. (1973) *Thermal Analysis*, Kogan Page, London, Ch. 5.
2. Linseis, M. (1975) *Proc. 4th ICTA*, Vol. 3, Heyden, London, p. 913.
3. Karmazsin, E., Satre, P. and Romand, M. (1982) *Proc. 7th ICTA*, Vol. 1, Wiley, Chichester, p. 337.
4. Levy, P. F. (1975) *Proc. 4th ICTA*, Vol. 3, Heyden, London, p. 3; (1971) *Int. Lab.*, Jan/Feb, 61.
5. Wendlandt, W. W. (1965) *Anal. Chim. Acta*, **33,** 98.
6. Gallagher, P. K. (1980) *Proc. 6th ICTA*, Vol. 1, Birkhaeuser, Basel, p. 13.
7. Schüller, K. H. and Kromer, H. (1982) *Proc. 7th ICTA*, Vol. 1, Wiley, Chichester, p. 526.
8. *Fisher Scientific Co., Bull.* 156, p. 4.
9. Hoffman, O. (1985) Proc. 8th ICTA, *Thermochim. Acta*, **93,** 529.
10. Shvaiko-Shvaikovsky, V. E. (1985) Proc. 8th ICTA, *Thermochim. Acta*, **93,** 493.
11. Hädrich, W., Kaiserberger, E. and Pfaffenberger, H. (1981) *Ind. Res. Dev.*, Oct., 165.

Chapter 7

Thermomechanical analysis (TMA)

7.1 Basic principles [1]

Thermodilatometry is carried out by measuring expansion and contraction under negligible loads. Further information of interest may be obtained by measuring the penetration, i.e. the expansion or contraction of a sample while under compression, as a function of temperature. Alternatively, the extension, i.e. the expansion or contraction of a sample under tension, may be measured as a function of temperature. These techniques, plus flexure and torsional measurements, are classified as thermomechanical analysis (TMA) and are obviously of great practical importance in materials testing (section 7.3).

7.2 Practical details

The apparatus used is based on that for thermodilatometry (chapter 6). The principles of penetration, extension, flexure and torsional measurements are illustrated in Fig. 7.1. Two main types of experiment are possible: (i) measurement of dilation with temperature at a fixed load, or (ii) measurement of dilation with load at a fixed temperature.

The Mettler TMA40 instrument is illustrated in Fig. 7.2. The overall measurement range is ±5 mm about the zero position set by the height adjustment ring. The probe force is microprocessor controlled with the help of a linear motor and ranges from 0 to 0.5 N. A dynamic force component can also be superimposed on the probe force during measurement (chapter 8).

The length is calibrated by means of gauge blocks of known size. The maximum sensitivity is ~4 nm. The temperature is calibrated using a special sandwich sample built up, as shown in Fig. 7.3, from discs of pure metals, separated by alumina discs. As the metals melt, the probe shows step-wise changes in length. The calibration sample can obviously only be used once. The instrument may be used from −100°C to 1000°C in atmospheres determined by the purge gas. The probe force is calibrated with a known mass.

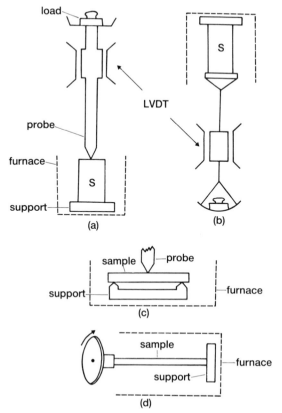

Figure 7.1 Thermomechanical analysis (TMA); (a) penetration, (b) extension, (c) flexure, and (d) torsional measurements.

Several different designs are available for the probe e.g. flat, pointed or rounded ends.

7.3 Applications of TMA

These techniques have been developed more recently than the others described so far, but applications have grown rapidly. TMA and TD (chapter 6) measurements are often carried out on the same sample with the same apparatus. Examples of both expansion and penetration measurements on neoprene rubber [2] are given in Fig. 7.4. The coefficient of linear thermal expansion, α, can be determined from the slope of the expansion curve (see chapter 6). Both the glass-transition (T_g) and melting (T_s) show up clearly. Another example [2] is the use of penetration measurements for testing

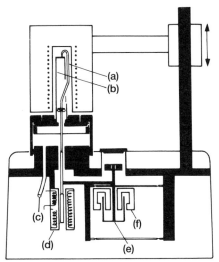

Cross section of the TMA40 thermomechanical analyzer.
(a) measuring sensor   (d) linear variable differential transformer
(b) sample support   (e) linear motor coil
(c) purge gas inlet   (f) linear motor stator

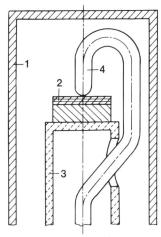

Measuring principle of the TMA:
1. metal furnace assembly with electric heating coil
2. sample (coated sheet metal)
3. sample holder (adjustable in height according to
   sample thickness)
4. measuring sensor with ball point (weight is compensated
   through weighing cell built in below,
   position is measured with linear transformer)

Figure 7.2 Mettler TA 3000 – Construction of the TMA40. (With the permission of Mettler Instrumente AG.)

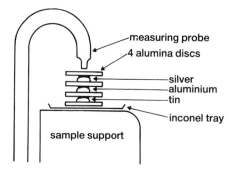

Figure 7.3 Temperature calibration for TMA. (With the permission of Mettler Instrumente AG.)

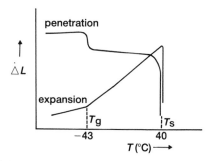

Figure 7.4 TMA and TD measurements on neoprene [2]. (With the permission of International Scientific Communications Inc.)

polyethylene-coated paper (Fig. 7.5). The structural transition and the melting show up clearly even though the coating was very thin ($<$0.3 mm). Similar measurements have been carried out on paint films [3] and can coatings. Blistering of coatings may be revealed [4].

The behaviour of polymer films and fibres on heating under load is of great practical importance. Figure 7.6 shows the behaviour of a polyimide film [7]. The film starts to shrink near 400°C. A similar measurement [4] on a drawn polyester fibre is shown in Fig. 7.7. Daniels [1] has given examples of torsional measurements.

Knowledge of the thermomechanical properties, especially the glass-transition temperatures, of polymers used as denture bases [5], is important as there are considerable variations in temperature in the mouth, as well as during the manufacture of dentures. Huggett *et al.* [5] have used TMA measurements to monitor the effect of various modifications of the polymerization of methyl methacrylate on the T_g of the resultant material. The factors expected [6] to increase the T_g of a polymer are:

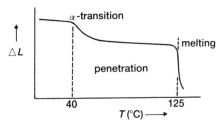

Figure 7.5 TMA measurements on polyethylene-coated paper [2]. (With the permission of International Scientific Communications, Inc.)

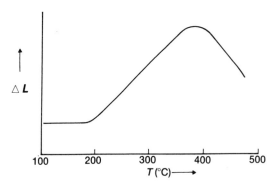

Figure 7.6 TMA measurements on polyimide film under load [7]. (With the permission of DuPont Analytical Instruments.)

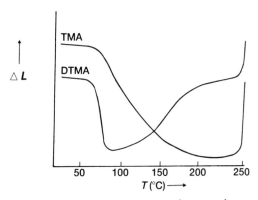

Figure 7.7 TMA and DTMA measurements on a drawn polyester fibre heated at 10 K min^{-1} in air, under load [4]. (With the permission of Stanton Redcroft Ltd.)

(i) The presence of groups in the backbone which increase the energy required for rotation, e.g. unsaturation and/or long side chains.
(ii) Secondary bonding between chains e.g. hydrogen bonding.
(iii) Crosslinking between chains.
(iv) Molecular mass.
(v) Copolymerization.
(vi) Plasticization.

7.4 References

1. Daniels, T. (1973) *Thermal Analysis*, Kogan Page, London, Ch. 5.
2. Levy, P. F. (1975) *Proc. 4th ICTA*, Vol. 3, Heyden, London, p. 3; (1971) *Int. Lab.*, Jan/Feb., 61.
3. Widmann, G. (1983) *Mettler Application No. 3401*.
4. Rumsey, J. A. and Ottoway, M. R. (1983) *Polymers Paint Colour Journal*, 23 Feb., p. 124.
5. Huggett, R., Brooks, S. C. and Bates, J. F. (1984) *Lab. Pract.*, **33** (11), 76.
6. Brydson, J. A. (1969) *Plastic Materials*, Iliffe Brooks Ltd., London.
7. Miller, G. W. and Casey, D. L. *DuPont Literature*.

Chapter 8

Dynamic mechanical analysis (DMA) or dynamic thermomechanometry

8.1 Basic principles [1, 2]

One of the techniques of thermomechanical analysis, described earlier in chapter 7, involves measurement of the change of a dimension of the sample as the load on the sample is increased, while the temperature is held constant. This is effectively the determination of a stress–strain curve for the sample. The load is related to the applied stress (force/area) and the change in dimension is related to the strain (elongation/original length).

Different types of materials give different types of stress–strain curves and, of course, the nature of the stress–strain curve for a given material will change as the temperature at which the measurements are made is changed.

8.1.1 Stress–strain curves

The two extremes of behaviour are elastic solids, which obey Hooke's law, i.e., that the applied stress is proportional to resultant strain (but is independent of the rate of strain), and liquids, which obey Newton's law, i.e., that the applied stress is proportional to rate of strain (not the strain itself). Both laws only hold for small values of strain or rate of strain.

8.1.2 Tensile stress

For a sample under tension the tensile stress, $\sigma = \text{force/area} = F/y_0 z_0$ and the tensile strain, $\varepsilon = \text{elongation/original length} = dx/x_0$. Hooke's law is $\sigma \propto \varepsilon$ or $\sigma = E\varepsilon$ where $E = $ Young's modulus (units: force/area). For an isotropic body

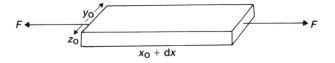

(i.e. properties independent of direction) the changes in length and in width are related by Poisson's ratio

$$v_p = -\left(\frac{dy}{y_0}\right)\left(\frac{x_0}{dx}\right) \qquad (v_p \text{ ranges from 0.2 to 0.5})$$

8.1.3 Shear stress

For a sample subjected to a shearing stress the shear stress, $\sigma_s = F/xy = F/A$, and the shear strain, $\varepsilon_s = dx/z = \tan\theta$. Again, Hooke's law is $\sigma_s \propto \varepsilon_s$ or $\sigma_s = G\varepsilon_s$, where G is the shear modulus. $G = \sigma_s/\varepsilon_s = F/A \tan\theta$ and for small θ, $\tan\theta \simeq \theta$ so $G \simeq F/A\theta$.

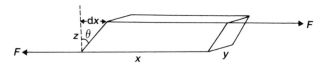

8.1.4 Compression

When a pressure p is applied to a body of original volume, V_0, causing a volume change, ΔV, the stress, $\sigma_B = p$, and the strain, $\varepsilon_B = -\Delta V/V_0$. Hooke's law is $\sigma_B \propto \varepsilon_B$ or $\sigma_B = B\varepsilon_B$, where B is the bulk modulus. $B = \sigma_B/\varepsilon_B = -pV_0/\Delta V$.

The compressibility of the sample is defined as $1/B = -\Delta V/pV_0$. Of the three moduli, E, G and B, only two are independent. They are related via Poisson's ratio

$$E = 3B(1 - 2v_p) = 2(1 + v_p)G$$

8.1.5 Newton's law for viscous liquids

The stresses applied to liquids are usually shear stresses and Newton's law is

$$\sigma_s \propto \frac{d\varepsilon_s}{dt}$$

(shear stress $\propto$ rate of shear strain) or

$$\sigma_s = \eta \frac{d\varepsilon_s}{dt}$$

where η is the viscosity coefficient.

8.1.6 Viscoelastic behaviour

Many polymers show behaviour intermediate between elastic solids and viscous liquids and such materials are classified as viscoelastic. Shear of viscoelastic materials is described by a combination of Hooke's and Newton's laws. Strains are additive, so when the total shear stress $=\sigma_s$,

$$\varepsilon_s = \varepsilon_{s,\text{elast}} + \varepsilon_{s,\text{visc}}$$

$$\varepsilon_{s,\text{elast}} = \frac{\sigma_s}{G} \text{ (Hooke's law) and } \frac{d\varepsilon_{s,\text{visc}}}{dt} = \frac{\sigma_s}{\eta} \text{ (Newton's law).}$$

Hence

$$\frac{d\varepsilon_{s,\text{elast}}}{dt} = \left(\frac{1}{G}\right)\frac{d\sigma_s}{dt}$$

so

$$\frac{d\varepsilon_s}{dt} = \frac{d\varepsilon_{s,\text{elast}}}{dt} + \frac{d\varepsilon_{s,\text{visc}}}{dt} = \left(\frac{1}{G}\right)\frac{d\sigma_s}{dt} + \frac{\sigma_s}{\eta}$$

When the shear strain is constant,

$$\frac{d\varepsilon_s}{dt} = 0 \quad \text{and} \quad \left(\frac{1}{G}\right)\frac{d\sigma_s}{dt} = -\frac{\sigma_s}{\eta}$$

which rearranges to

$$\frac{d\sigma_s}{\sigma_s} = -\frac{G}{\eta}dt$$

If $\sigma_s = \sigma_0$ at $t=0$

and $\sigma_s = \sigma$ at $t=t$

$$\int_{\sigma_0}^{\sigma} \frac{d\sigma_s}{\sigma_s} = -\frac{G}{\eta}t$$

i.e. $\ln\left(\dfrac{\sigma}{\sigma_0}\right) = -\dfrac{G}{\eta}t \quad \text{or} \quad \sigma = \sigma_0 \exp\left(-\dfrac{G}{\eta}t\right)$

This means that at a fixed strain, if the initial stress is σ_0, this will relax exponentially with time. The relaxation time is η/G.

8.1.7 Stress–strain measurements on polymers

If a viscoelastic sample is subjected to a tensile force, applied at a uniform rate, and the resultant elongation of the sample is measured, a curve of the form shown in Fig. 8.1(a) may be obtained. (The detailed shape will depend upon the rate of application of the stress.) In the region OL, Hooke's law is obeyed and the slope gives Young's modulus, E. Beyond the yield point, Y, viscous flow occurs until the maximum elongation is achieved at the break point, B.

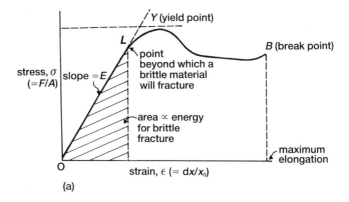

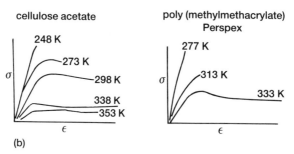

Figure 8.1 (a) Typical stress–strain curve for a viscoelastic sample [1]. (b) Effect of temperature on stress–strain curves [1]. (Based on Cowie, J. M. G. (1973) *Polymers: Chemistry and Physics of Modern Materials*, Intertext, New York., Ch. 12, with permission of Blackie and Son Ltd., Glasgow.)

The effects of temperature on the stress–strain curves are illustrated in Fig. 8.1(b). As the temperature increases (i) the modulus of elasticity, E, (i.e. slope of OL in Fig. 8.1(a)) decreases; (ii) the yield strength decreases; and (iii) the maximum elongation generally increases.

8.1.8 Periodic stresses

In dynamic mechanical analysis (DMA) the sample is subjected to a sinusoidally varying stress of angular frequency, ω. For a viscoelastic material, the resulting strain will also be sinusoidal, but will be out of phase with the applied stress owing to energy dissipation as heat, or damping, in the sample (Fig. 8.2(a)). δ is the phase angle between the stress and the strain.

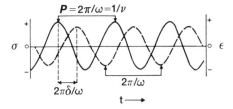

(a) Forced oscillation [1]

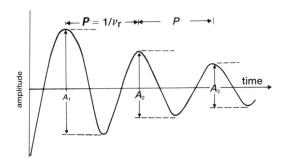

(b) Free oscillation with damping [1]

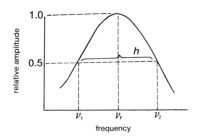

(c) Measurement of damping [1]

Figure 8.2 (Based on Cowie, J. M. G. (1973) *Polymers: Chemistry and Physics of Modern Materials*, Intertext, New York., Ch. 12, with permission of Blackie and Son Ltd., Glasgow.)

8.1.9 The resonance frequency, v_r

If a sample is vibrated over a range of frequencies and the amplitude of vibration is measured, the resonance frequency is that which produces a maximum in a plot of amplitude against frequency. Young's modulus (of elasticity), E, is related to the square of the resonance frequency, v_r.

$$E = c\, L^4 \rho v_r^2 / B^2$$

where c is a constant, L is the sample length between clamps, B is the sample thickness and ρ is the sample density. If the sample is allowed to oscillate freely, it will do so at the natural resonance frequency with decreasing amplitude owing to damping. The form of the oscillations is shown in Fig. 8.2(b). The period, P, of the oscillations is constant $(=1/v_r)$ and the amplitude (measured from a minimum to the preceding maximum) decays exponentially. The rate of decay is a measure of how much damping there is. Damping is usually expressed as the logarithmic decrement per cycle, $\Delta = \log_{10}(A_1/A_2) = \log_{10}(A_2/A_3) = \ldots$ etc. Units used are decibels (dB) i.e. $10 \times \log(A_1/A_2)$, e.g. if the amplitude halves, $\log(A_1/A_2) = \log 2 = 0.30$. Damping is then 3.0 dB.

Damping may be determined from the width of a curve of relative amplitude against frequency (Fig. 8.2(c)), $h = (v_2 - v_1)/v_r$, or by measuring the driving force required to maintain a constant amplitude of vibration at the resonance frequency (forced vibration).

Damping depends on the physical state of the sample, e.g. at $T > T_g$, polymers dissipate most of the energy supplied and damping is high. At $T < T_g$, the material stores the energy and damping is low.

8.1.10 DMA apparatus

The DuPont DMA instrument is illustrated in Fig. 8.3. The sample, which must be in film or fibre form, is clamped between the arms of the complex pendulum. The arms-plus-sample system is set in motion by the electromechanical

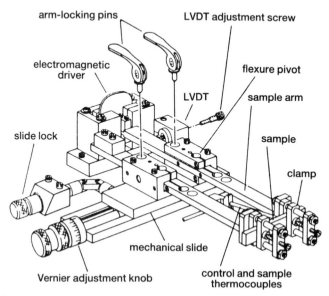

Figure 8.3 DuPont 982 Dynamic Mechanical Analyser. (With permission of DuPont Analytical Instruments.)

transducer. A linear variable differential transformer (LVDT; see Fig. 6.2) is used to monitor the frequency and the amplitude of vibration and feedback from the LVDT is used to control the electromechanical transducer. The preset oscillation amplitude can then be maintained and the driving force required to do so is a measure of the energy dissipation (damping) of the sample. The DMA outputs are thus plots of resonance frequency and of damping as functions of temperature. Young's modulus, E, for the sample is obtained from the resonance frequency, v_r

$$E = \frac{(4\pi^2 \, v_r^2 \, J_0 - K)}{2W\left[\left(\dfrac{L}{2}\right)+D\right]^2} \left(\frac{L}{B}\right)^3$$

where J_0 is the moment of inertia of the arm system, K is the spring constant of the pivot, D is the clamping distance, W is the sample width, B is the sample thickness, and L is the sample length.

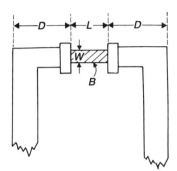

Connop *et al.* [3] have made a careful study of the critical factors in using the DuPont 981 DMA, and have shown that to obtain reproducible results it is necessary to: (i) keep all sample dimensions within an aspect ratio of from 14 to 20; (ii) mount the sample consistently with careful centering and tightening of the clamps; (iii) maintain a constant gas flow-rate; (iv) maintain a constant sample thermocouple position.

Although samples should ideally be rigid enough to be tested as a sheet or rod, softer materials, such as thermoplastics or elastomers can be studied by using specially designed supports [4]. With thin films, buckling deformation can be a problem. Toth *et al.* [5] have developed a special sample clamp assembly which provides an arched sample cross-section. Miller [6] describes a technique for characterizing organic coatings by DMA by coating a multiple filament glass strand. The contribution of the substrate alone is subtracted to give the contribution of the coating. A similar technique using a thin steel strip as substrate can be used in examining the mechanical properties of paints.

The polymer laboratories instrument (Fig. 8.4) normally operates with a bar

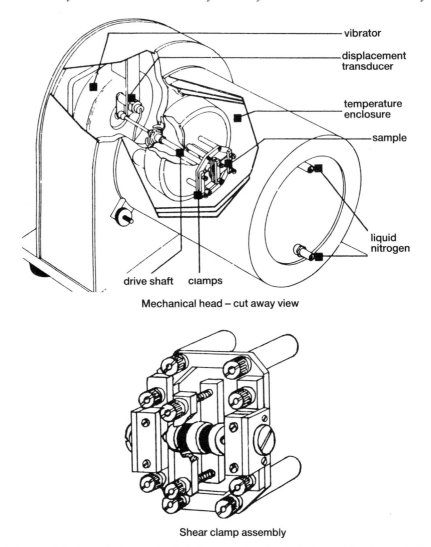

Mechanical head – cut away view

Shear clamp assembly

Figure 8.4 Polymer Laboratories Ltd. Dynamic Mechanical Thermal Analyzer. (With the permission of Polymer Laboratories Ltd.)

sample which is clamped at both ends. The central point of the bar is vibrated sinusoidally by the drive clamp on the end of a ceramic drive shaft. The stress experienced by the sample is proportional to the current supplied to the vibrator and the strain, which is proportional to the sample displacement, is monitored by the transducer. Softer materials are measured in the shear clamp assembly shown and liquid samples can be supported on films or absorbed into papers or braids.

8.2 Applications of DMA

Changes in Young's modulus indicate changes in rigidity and hence strength of the sample. Damping measurements give practical information on glass transitions, changes in crystallinity, the occurrence of cross-linking, and also show up the features of polymer chains. The information obtained has been used in various very practical areas such as studies of vibration dissipation, impact resistance and noise abatement.

Typical DMA results [7] on two different samples of polyethylene are shown in Fig. 8.5(a) and (b). The damping curve for linear polyethylene (Fig. 8.5(a)) shows peaks at $-95°C$ and $65°C$. The lower temperature peak has been attributed to long chain $(-CH_2-)_n$ crankshaft relaxations in the amorphous

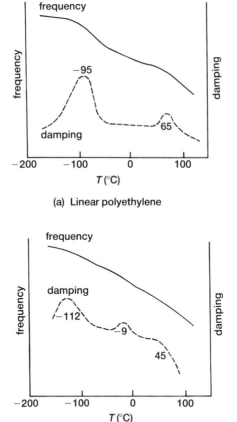

(a) Linear polyethylene

(b) Branched polyethylene

Figure 8.5 DMA curves of polyethylene samples [7]. (With the permission of DuPont Analytical Instruments.)

phase and the higher temperature peak to similar motion in the crystalline phase. The temperatures and relative sizes of the two peaks can be related to the degree of crystallinity of the sample. The damping curve for branched polyethylene (Fig.8.5(b)) has features at $-112°C$, $-9°C$ and $45°C$. The $-112°C$ and $45°C$ peaks are explained as above, while the $-9°C$ peak is attributed to $-CH_3$ relaxations in the amorphous phase.

In Fig. 8.6, the behaviour of styrene–butadiene-rubber (SBR) has been examined [7, 8]. Various formulations of SBR are used in tyre manufacture. Different styrene–butadiene ratios may be used, or different butadiene isomers,

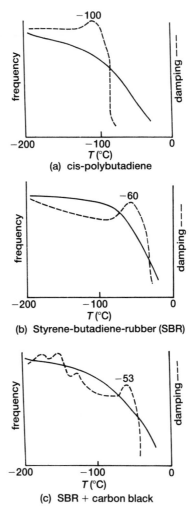

Figure 8.6 DMA curves on rubber blends [7, 8]. (With permission of DuPont Analytical Instruments.)

or different additives, e.g. carbon black. A high *cis*-butadiene content lowers the glass transition temperature, T_g (to as much as $-110°C$ compared to $-50°C$) giving greater flexibility at low temperatures. The addition of carbon black (Fig. 8.6(c)) increases the modulus of elasticity. The T_g value is also slightly increased. The complex damping curve at low temperatures indicates

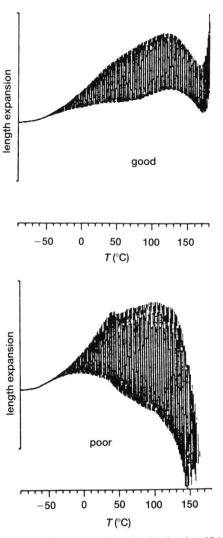

Figure 8.7 DMA curves of a well and poorly-vulcanized polysulfide product, where the plastic flow begins at a lower temperature due to light crosslinking. Heating rate 10 K min^{-1}, temperature range -90 to $180°C$. (Mettler Application Digest, with permission.)

polymer–carbon black interactions and may lead to adverse properties, e.g. heat build-up.

Huson *et al.* [4] have used the changes in height of DMA damping peaks as a measure of the extent of vulcanization in elastomer blends. Results agreed well with torque rheometer measurements.

Gill [9] has given a general review of the application of DMA to polymer composites.

Tóth *et al.* [5] have used DMA to study the effect of ageing on paper. Samples were subjected to accelerated ageing at 100°C in air with 50% relative humidity and the DMA curves of original and aged samples were recorded. The increased brittleness of the aged samples showed up in the damping curves.

Schenz *et al.* [10] have designed a sample mould for preparing frozen aqueous solutions for examination by DMA. The DMA technique was found to be very sensitive to second-order transitions. Frozen dilute solutions of sucrose showed two glass transitions at around $-32°C$, suggesting that as the solutions are frozen, tiny localized inclusions are formed where the concentration differs from the bulk concentration. These inclusions have a T_g higher than that of the bulk solution. DSC was not sensitive enough to detect these details.

TMA instruments (chapter 7) may be capable of imposing a periodic stress on the sample. An example of such a measurement made on the Mettler TMA40 (Fig. 7.2) is shown in Fig. 8.7. The extent of vulcanization of two polysulphide samples is compared. The load alternates with cycle time of 12 s.

Kaiserberger [11] has compared the performance of TMA, DMA and DSC for determining the viscoelastic properties of polymers and concludes that DMA has superior sensitivity in detecting phase transitions of second and higher order.

8.3 References

1. Cowie, J. M. G. (1973) *Polymers: Chemistry and Physics of Modern Materials*, Intertext, New York, Ch. 12.
2. Ward, I. M. (1971) *Mechanical Properties of Solid Polymers*, Wiley, New York.
3. Connop, A., Huson, M. G. and McGill, W. J. (1982) *J. Thermal Anal.*, **24**, 223.
4. Huson, M. G., McGill, W. J. and Swart, P. J. (1984) *J. Polym. Sci., Polym. Lett.*, **22**, 143.
5. Tóth, F. H., Pokol, G., Gyore, J. and Gal, S. (1985) Proc. 8th ICTA, *Thermochim. Acta*, **93**, 405; (1984) **80**, 281.
6. Miller, D. G. (1982) *Int. Lab.*, March, 64.
7. DuPont Instruments, *Thermal Analysis Review, Dynamic Mechanical Analysis*, undated.
8. Hassel, R. L., *DuPont Application Brief*, TA68.
9. Gill, P. S. (1983) *Ind. Res. Dev.*, March, 104.
10. Schenz, T. W., Rosolen, M. A., Levine, H. and Slade, L. (1984) *Proc. 13th NATAS Conf.*, paper 12, p. 57.
11. Kaiserberger, E. (1985) Proc. 8th ICTA, *Thermochim. Acta*, **93**, 291.

Chapter 9

Combination of thermal analysis techniques

9.1 Principles

Results of the thermal analysis of a given sample, under a specified set of conditions, on a given instrument, are usually reasonably reproducible. Agreement with the results on another portion of the same sample, obtained using the same technique on an instrument of a different make, may be much less satisfactory. It is even more difficult to compare results of parallel measurements from two or more independent thermal analysis techniques, e.g. TG and DTA. The advantages of being able to relate results of different techniques are obvious. For example TG cannot be used to detect melting, while melting and decomposition cannot be distinguished unambiguously using DTA. A substance which melts with accompanying decomposition must thus be studied using both TG and DTA (or DSC). There is usually a synergistic effect in that the total amount of information about the sample that is obtained is greater than the sum of the information obtained from the individual techniques.

A truly simultaneous technique involves measurements of two or more of the properties in Table 1.1 on the same portion of the sample during a single temperature programme. Each of the properties may be recorded continuously, or they may be sampled in a repetitive sequence to allow, for example, the data to be captured by computer (chapter 12). Simultaneous measurements must thus be distinguished from parallel measurements, where different portions of the sample are examined using different instruments, and concurrent measurements where different portions of the same sample, in different containers, are held within a single furnace and are subjected to a common temperature programme (Fig. 9.1). This last option is also used in instruments designed for carrying out one type of measurement on several samples.

Simultaneous measurements obviously take less time than sets of separate measurements, but generally the sensitivity of the individual techniques is reduced on combination, because of compromises in instrumental design. In

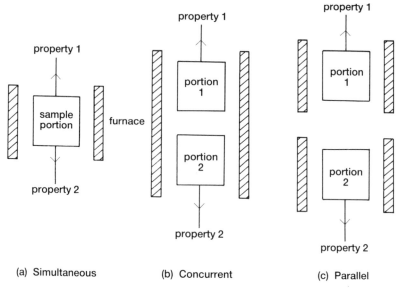

(a) Simultaneous   (b) Concurrent   (c) Parallel

Figure 9.1 Combinations of thermal analysis techniques.

considering simultaneous methods, two combinations which probably fit the above definition are not considered further. These are the simultaneous recording of derivative curves of the basic signal, e.g. TG and DTG, and the combination of evolved gas analysis (EGA) with any other thermal analysis technique. EGA is discussed in detail in chapter 10 and its effect on the sensitivity of the basic technique is usually minimal.

Leaders in the use of simultaneous methods, based on their instrument called the 'Derivatograph', have been J. and F. Paulik [1]. The Derivatograph (Fig. 9.2) is an example of one of the most successful combinations, i.e. that of TG with DTA to give TG-DTA. In the Derivatograph and some other instruments, the DTA reference material is not weighed. In the Stanton Redcroft TG-DTA system (Fig. 9.3) [2] the whole DTA configuration together with its thermocouple leads is incorporated into the balance suspension. A different approach is used in the Setaram simultaneous TG-DSC 111 (Fig. 9.4) [3]. There is no mechanical contact of the TG sample pans with the detectors of the DSC tubes. This is claimed not to affect the quantitative performance of the detectors. The limit of detection of the DSC is given as 30 μW in scanning mode. The symmetrical configuration of the system and the absence of contact problems are both factors in favour of accurate weighing (chapter 3). Demensky *et al.* [4] used a very similar system based on a Setaram DSC 111 and reported that the time constant of the calorimeter was $\sim$ 20 s and that the DSC curve was thus distorted. They used a Fast Fourier transform method to obtain true DSC

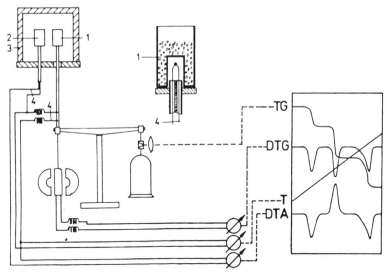

Figure 9.2 J. and F. Paulik's Derivatograph [1] for simultaneous TG-DTA. 1. Sample 2. Reference 3. Furnace 4. Thermocouples. (With the permission of Elsevier, Amsterdam.)

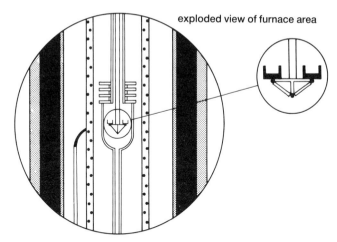

Figure 9.3 The sample holder assembly of the Stanton Redcroft STA-780 simultaneous TG-DTA system [2]. (With the permission of Stanton Redcroft Ltd.)

curves. Gast *et al.* [5] have described a magnetically coupled thermobalance adapted for simultaneous DTA.

Because of the sophistication of the principle involved, combination of power-compensated DSC (chapter 4) with another technique such as TG is not likely to be achieved.

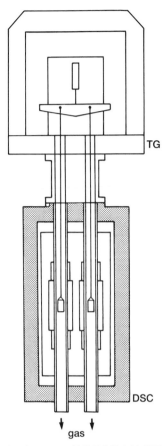

Figure 9.4 The Setaram simultaneous TG-DSC 111 [3]. (With the permission of Setaram Ltd.)

DTA may be combined with TMA by inserting thermocouples in the sample and in the reference. DTA has been successfully combined with hot-stage microscopy (chapter 5) and even small-angle X-ray scattering [6]. Many of the less-common properties, whose measurement is described in chapter 11, have been measured simultaneously with more conventional properties.

An example of a concurrent measurement system is the modification of the DuPont 951 TG system, described by Mikhail [7], in which a second thermocouple was added to that already used for the TG, and sample and reference pans were placed in contact with the junctions of the two thermocouples. DTA measurements were then carried out concurrently with TG on a separate portion of the sample.

Johnson and Ivansons [8] have described a DuPont multiple-sample DSC which can run up to three samples concurrently. It is necessary that all positions for the samples and the reference should be thermally equivalent and that there should be no thermal communication (crosstalk) between the positions. Such a multiple-sample system, in addition to allowing for increased throughput of samples, also allows for inclusion of reference materials for calibration concurrently with the samples.

While dealing with simultaneous measurements, it is worth stressing the importance of complementary techniques which can aid in identifying the thermal events occurring in the sample. Complementary techniques include all forms of spectroscopy, X-ray and electron diffraction, optical and electron microscopy and are limited only by the ingenuity of the investigator.

9.2 References

1. Paulik, J. and Paulik, F. (1981) Simultaneous Thermoanalytical Examinations by means of the Derivatograph, *Wilson and Wilson's Comprehensive Analytical Chemistry*, Vol. XIIA, Elsevier, Amsterdam.
2. Charsley, E. L., Joannou, J., Kamp, A. C. F., Ottaway, M. R. and Redfern, J. P. (1980) *Proc. 6th ICTA*, Vol. 1, Birkhaeuser, Basel, p. 237.
3. Le Parlouër, P. (1985) Proc. 8th ICTA, *Thermochim. Acta*, **92,** 371.
4. Demensky, G. K., Petrov, L. A., Tarasenko, Y. V. and Teplov, O. A. (1985) Proc. 8th ICTA, *Thermochim. Acta*, **92,** 265.
5. Gast, Th., Jakobs, H. and Mirahmadi, A. (1981) *Proc. ESTA 2*, Heyden, London, p. 3.
6. Russell, T. P. and Koberstein, J. P. (1984) *Proc. 13th NATAS Conf.*, paper 104, p. 443.
7. Mikhail, S. A. (1985) *Thermochim. Acta*, **95,** 287.
8. Johnson, R. C. and Ivansons, V. (1981) *Proc. 11th NATAS Conf.*, paper 40, p. 237.

Chapter 10

Evolved gas analysis (EGA)

10.1 Basic principles

Many samples, on heating, release gases or vapour through desorption or decomposition. This release is accompanied by thermal effects and, obviously, mass losses, which, themselves, can be detected by the appropriate thermal analysis technique, e.g. DTA or DSC and TG respectively. The thermal analysis technique does not, however, identify the gas evolved and, for complex decompositions, such information is essential. It has thus become fairly routine to couple the basic techniques already described with a system for either detecting the evolution of gas (or gases) from the sample (**evolved gas detection, EGD**) or, more satisfactorily, detecting and identifying the gases evolved (**evolved gas analysis, EGA**) [1]. The apparatus for EGA will obviously be more complex than that required for EGD.

10.2 Evolved gas detection (EGD)

EGD has the advantage that measurements may be continuous and hence are readily related to thermal analysis curves. In most thermal analysis techniques, a purge gas is used and this then becomes the carrier for sweeping evolved gases to a detector (Fig. 10.1). The detector should be as close as possible to the sample to reduce condensation of vapours, secondary reactions in the gas phase, and time lags between thermal analysis and EGD curves. The most commonly used detectors [1] are those usually found in simpler gas chromatographs, namely: (i) thermal conductivity detectors (TCD, or katharometers); (ii) gas-density detectors; and (iii) ionization detectors. In addition, use has been made of infrared radiometers.

For maximum sensitivity the measured property (e.g. thermal conductivity) of the evolved gas should differ markedly from that of the carrier (Table 10.1). H_2 and He have very high, and argon very low, thermal conductivities, making them suitable carriers, but their influence on the thermal effects being examined, must be determined.

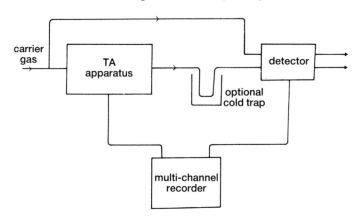

Figure 10.1 Evolved gas detection (EGD).

Table 10.1 Thermal conductivities of some gases

Gas	Thermal conductivity, λ (Wm^{-1} K^{-1})		Differences, $100\ \Delta\lambda = 100(\lambda_{gas} - \lambda_{ref})$		
	$T = 300$ K	$T = 373$ K	Ref. N_2	Ref. Ar	Ref. He
Air	0.0262	0.0301	+0.01	+0.84	−12.47
Ar	0.0178	0.0213	−0.83		−13.31
He	0.1509	0.1669	+12.48	+13.31	
H_2	0.1869	0.2090	+16.08	+16.91	+3.60
N_2	0.0261	0.0301		+0.83	−12.48
O_2	0.0266	0.0311	+0.05	+0.88	−12.43
CO	0.0251	0.0305	−0.10	+0.73	−12.58
CO_2	0.0166	0.0212	−0.95	−0.12	−13.43
NH_3	0.0246	0.0327	−0.15	+0.68	−12.63
N_2O	0.0174	0.0212	−0.87	−0.04	−13.35
NO	0.0260		−0.01	+0.82	−12.49
NO_2	0.0372 (328 K)		+1.11	+1.94	−11.37
SO_2		0.0153 (est.)	−1.1	−0.3	−13.6
H_2O(g)	0.0181	0.0241	−0.80	+0.03	−13.28

Sources: Weast, R. C. (ed.) (1984) *Handbook of Chemistry and Physics*, CRC Press, Boca Raton, 64th Edn, p. E-2. Lange, N. A. (ed.) (1961) *Handbook of Chemistry*, McGraw-Hill, New York, 10th Edn, p. 1544. Lodding, W. (ed.) (1967) *Gas Effluent Analysis*, Edward Arnold, London, p. 30. Thomas, L. C. (1980) *Fundamentals of Heat Transfer*, Prentice-Hall, Englewood Cliffs, p. 665.

Flame ionization detectors are particularly sensitive to organic compounds, but do not respond to water vapour.

Some separation of gas mixtures is possible by carrying out several EGD runs with suitable cold traps interposed between the sample and the detector. Alternatively, two similar detectors may be used, one on either side of the trap.

10.3 Evolved gas analysis (EGA)

10.3.1 Mass spectrometry (MS)

The most versatile and fastest means of repetitive gas analysis is mass spectrometry, so the most obvious solution to the problem of identifying the gases evolved from a thermal analysis instrument, is to replace the detector (Fig. 10.1) by a mass spectrometer [2, 3]. Such a replacement is, however, accompanied by several problems. Mass spectrometers are expensive items of equipment and dedicated use on a thermal analysis instrument is unlikely to be received enthusiastically by other potential users. The development of quadrupole mass spectrometers (QMS) at somewhat lower prices has eased this problem and Emmerich and Kaiserberger [4] have suggested that the capital already invested in the thermal analysis equipment warrants the further investment to get the maximum information per run. A second problem is that mass spectrometers require high vacuum for their operation, while most thermal analysis experiments are carried out at, or just above, atmospheric pressure. Some attempts have been made [2, 3] to carry out the thermal analysis experiments inside the vacuum system of the MS (Fig. 10.2(a)). Results obtained on heating a sample in vacuum are, however, often very difficult to relate to the behaviour of the sample under atmospheric pressure. It is thus more usual [4] to use some form of molecular leak inlet to introduce samples of the carrier gas and the accompanying evolved gases from the thermal analysis instrument into the MS (Fig. 10.2(b)).

Problems that can arise include: (i) condensation of vapours in the sampling system – attempts to reduce this by heating the system promote secondary reactions; (ii) the high concentration of carrier gas may swamp the smaller responses of evolved gases – He, with its low mass number, is thus useful as a carrier.

An alternative to both of the above systems is to use a jet separator in the MS inlet (Fig. 10.2(c)). The jet separator allows the evolved gases to be separated from helium as carrier before introduction into the MS [5–7]. Dyszel [8] has described the interfacing of a TG system to an MS designed for atmospheric pressure chemical ionization (APCI). Gases entering the APCI source are ionized in a discharge corona. Only selected ions are then admitted to the MS through an electrostatically controlled inlet system.

Complete mass spectra may be recorded repetitively, or selected mass numbers may be sampled using a suitably programmed system, or a single mass number may be monitored continuously.

In interpreting the results of MS studies, allowance has to be made for the fragmentation patterns of the parent product molecules. Excessive fragmentation of the gas molecules occurs in some cheaper quadrupole mass spectrometers which do not have a variable ionization potential. Mixtures of CO_2 and CO, for example, then appear mainly as C^+ and O^+ fragments.

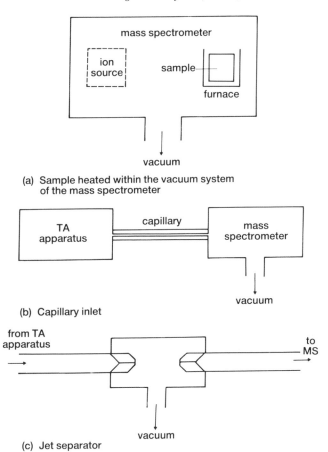

Figure 10.2 Coupling thermal analysis equipment to a mass spectrometer.

10.3.2 Gas chromatography (GC)

Another approach to EGA is to precede the detector (Fig. 10.1) by a column of a suitable adsorbent, i.e. to pass the evolved gases from the thermal analysis through a gas chromatograph [9–11]. Sampling, of course, now becomes intermittent as time has to be allowed for the component of the gas mixture with the longest retention time to be eluted from the column, before the next sample is introduced. This is the main disadvantage of GC, since times required for adequate separations may be of the order of several minutes. This means that only a few samplings may be possible during a rapid thermal event, and such events may even be missed completely. The advantage of GC is, however, that by suitable choice of column-packing material, most separations can be achieved and retention times, once determined, provide a simple means of identification. The vast literature of gas–solid chromatography is available in deciding upon a suitable set of analysis conditions for a given sample and its expected products. Two useful references on the analysis of mixtures of the more commonly-encountered gases are the books by Hachenburg [12] and Thompson [13]. The ultimate, of course, is to separate the evolved gases on a column and to confirm their identification through use of an MS as the GC detector. The gas chromatograph (which may be assembled from selected components) has to include a gas-sampling value (Fig. 10.3) operated by a programmable timer.

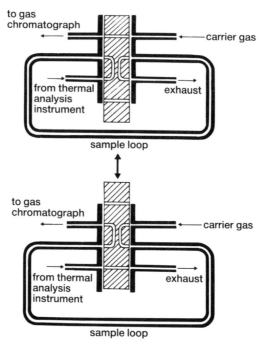

Figure 10.3 Gas-sampling valve (Taylor Servomex). (With the permission of Servomex Ltd.)

10.3.3 Infrared spectroscopy

Morgan [14] has described the use of non-dispersive (i.e. fixed wavelength) infrared analyzers, coupled in series to a DTA apparatus, to measure the amounts of CO_2 and H_2O evolved by minerals and mineral mixtures. SO_2 is another common gaseous decomposition product and this was determined with a special electrolytic cell (see below) also in series. The infrared analyzers are shown schematically in Fig. 10.4. Two intermittent beams of infrared radiation of equal energy are passed simultaneously through the sample and reference

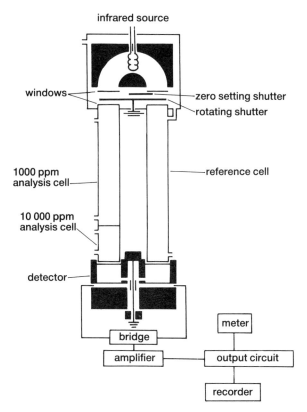

Figure 10.4 Non-dispersive infrared analyzer [14]. (With the permission of Wiley–Heyden Ltd.)

cells and into a detector. The detector consists of two symmetrical chambers, divided by a pressure-sensitive diaphragm and filled with a sample of the type of gas to be measured. The reference cell is filled with a non-absorbing gas. When the sample gas is passed through either of the analysis cells (depending upon the concentration range) it absorbs energy and this imbalance causes deflection of the detector diaphragm. The movement of the diaphragm is converted to an electrical signal. The evolution profiles are integrated and the apparatus is calibrated with standard gas mixtures. A relatively fast flow rate of carrier gas is used and the delay in the EGA signal for CO_2 and SO_2 relative to the DTA signal was found to be negligible (< 5 s) with a longer delay (30 s) for H_2O caused by temporary condensation between furnace and analysis cell.

Both inert and oxidizing atmospheres can be used so that solid–gas interactions and combustion reactions can be studied. Filsinger and Bourrie [15] have used a very similar system consisting of a Netzsch thermal analyzer with a Lira Model 202 IR analyzer for CO, to study the gases evolved in the decomposition of calcium oxalate. It was thought that the decomposition of CaC_2O_4 to $CaCO_3$ would be useful as a calibration standard for CO, but the disproportionation of the CO turned out to be significant at the temperatures used, even though thermodynamic equilibrium was not reached.

The use of Fourier transform infrared spectrometry (FT-IR) for evolved gas analysis has been reviewed by Lephardt [16]. The potential use of FT-IR for the identification of products of thermal degradation had been suggested by Liebman *et al.* [17], who remarked on the specificity and the short measurement times required, as well as warning of the possibilities for interference from the rotational–vibrational fine structure of spectra of small molecules such as H_2O and HCl.

In considering the apparatus required [16] for FT-IR-EGA one is faced with some of the general problems discussed in chapter 9 on simultaneous methods. Compromises have to be made between the ideal requirements for operation of individual instruments, e.g. high heating rates and low carrier-gas flow rates will give greater concentrations of degradation products in the analysis cell, but these conditions are not always suitable for accurate thermal analysis. Secondary reactions between gaseous products are also enhanced at low flow rates. Lephardt [16] deals with both the coupling of commercially available TA equipment to FT-IR instrumentation and the development of specialized thermal units for FT-IR-EGA.

Liebman *et al.* [17] used a flow-through, heated light-pipe gas cell (30 cm long and 4×4 mm cross section). Collimated IR radiation was passed into an interferometer and the exit beam from the interferometer was focused on to the window of the cell. The beam emerging from the light-pipe was focused on to the IR detector. A predetermined number of interferograms (usually 50), were signal averaged and each set was stored for later recall and transformation. The resulting spectra were ratioed against a single-beam spectrum of the background. Spectra were measured at 4 cm^{-1} resolution. The time interval between spectra, for 50 scan sets, was 2 min, but this could be reduced by reducing the number of scans. This increased the noise, especially above 2500 cm^{-1}. The spectra obtained [17] showed that FT-IR allows all volatile products to be monitored simultaneously.

Lephardt and Fenner [18] ensured that interferograms were accurately associated with sample temperatures. They also arranged for transfer of large arrays of spectra from the spectrometer to a secondary computing system, so that the dedicated computer could continue the data capture, while the larger computer did the data processing. Output was available as absorbance against temperature plots. The system was applied to the pyrolysis of tobacco [18]. Later developments [19] included standard routines for smoothing (Savitsky–Golay) and integration (trapezoidal rule) (see chapter 12) as well as more advanced processing of the EGA interferograms, using comparison with interferograms of compounds of interest, to deconvolute overlapping profiles into their components. In addition to compound-specific profiles, reaction-specific profiles could be produced.

Davidson and Mathys [20] describe the use of a ratio recording dispersive infrared spectrometer for EGA. Gas cells of from 10 cm to multipass cells of 1 m path length and low volume (45 ml) were used and digitized spectra at a resolution of 5 cm^{-1} were recorded in a scan time of 1 to 2 min and smoothed. Heating rates of 5 K min^{-1} were used to allow adequate definition of gas evolution profiles. Most effluents yield relatively simple spectra, and evolution profiles were constructed from measurements on selected absorption bands. The temperature corresponding to each measured absorption was determined from the scan time and the heating rate, with allowance for times required for writing data to disk. The gas transfer lines and sample cell were at the temperature of the spectrometer (32°C) and water condensation problems were not experienced at the low levels used. The pyrolysate from some polymers, e.g. polyethylene, could cause contamination problems.

10.3.4 Special-purpose detectors

There is no reason, in principle, why any means of analyzing gases should not be coupled to a thermal analysis instrument, and there are reports of the use of absorbents and volumetric methods for the determination of total amounts of evolved gases. Detailed information on the evolution of each component with time is, however, most desirable.

Water vapour is a very common product of thermal decompositions, as well as being a product of the reduction of metal oxides with hydrogen, and special detectors have been developed to monitor evolution of water, in addition to the use of non-dispersive infrared analyzers described above [14]. Many of the common methods for determining the water content of gases are not entirely suitable when the water content is varying fairly rapidly, on account of slow and/or non-linear detector responses. For EGA the response of the detector should also be selective for water as other products may be evolved simultaneously.

Warrington and Barnes [21] have shown that an electrolytic hygrometer is suitable for continuous water analyses. The hygrometer has two fine platinum wires wound closely, but not in contact, on a PTFE former. The wires are coated with phosphoric acid and the whole element is enclosed in a flow-through glass tube which may be coupled to the outlet of a TA apparatus. At the start, the acid is electrolyzed to dryness using a potential of 100 V between the two wires, and, in this state, there is then a negligible current between the two wires. The acid coating absorbs any water from the gas stream passing through the tube and this water is then electrolyzed. The electrolysis current is proportional to the water concentration in the gas. Precautions necessary to avoid spurious effects are described [21], and the system was tested on the dehydration of several hydrates and the decompositions of several carboxylic acids.

Gallagher *et al.* [22, 23] have developed an EGA system based on a Panametrics Model 700 Moisture Analyser in which the moisture content is determined from the dew point of the gas stream and the flow rate. The relationship is non-linear and the computations are described [22]. Allowance has to be made for background moisture in the gas stream and for degassing of the TA apparatus as the temperature rises, as well as loss of moisture by adsorption on cooler surfaces of the system.

Morgan [14] has described an electrolytic detector for SO_2 analysis (Fig. 10.5) based on a fuel cell. SO_2 in the carrier gas diffuses through a semi-permeable membrane and is adsorbed on a sensing electrode, producing a current in the circuit which is proportional to the partial pressure of SO_2 in the carrier gas. Calibration using a standard gas mixture is required.

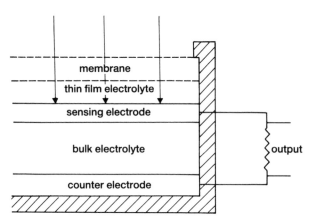

Figure 10.5 Electrolytic cell for SO_2 analysis [14]. (With the permission of Wiley–Heyden Ltd.)

Gauthier *et al.* [24] have discussed the use of solid electrolytes of K_2SO_4 for analysis of SO_2 and K_2CO_3 for CO_2. The latter has been used in a study of the decomposition of $CaCO_3$.

A review in *Analytical Chemistry* [25] points to some of the revolutionary new sensors under development, some of which will undoubtedly find application in EGA systems of the future.

Chemical conversion agents have been suggested to simplify the final analysis, e.g. use of I_2O_5 to convert CO to CO_2 and of CaC_2 to convert H_2O to C_2H_2 for easier GC analysis. The kinetic aspects of these processes introduce undesirable uncertainties in EGA.

10.4 Applications of EGD and EGA

The main use of EGD is to distinguish between phase transitions and endothermic decompositions. This is of particular use in studies on coordination compounds, where loss of ligand is usually endothermic and many structural changes, including melting, are possible.

In Fig. 10.6(a), EGD curves using a thermal conductivity detector with and without a preceding liquid air (90 K) trap, and recorded simultaneously with the DSC trace, are shown for the decomposition of nickel oxalate dihydrate. These responses were interpreted as dehydration followed by a complex endothermic decomposition with evolution of mainly CO_2 and traces of CO (not condensed at liquid air temperature).

A closely-related study on the isothermal decomposition of nickel formate dihydrate in the DSC at 480 K with simultaneous EGA by GC is illustrated in Fig. 10.6(b). There is concurrent evolution of CO_2, CO and H_2O. The relative proportions would have to be determined by calibration e.g. by dehydrating a

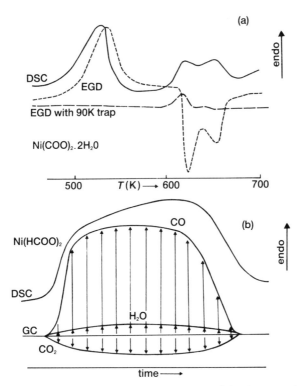

Figure 10.6 EGA coupled with DSC used in the study of the decomposition of metal carboxylates. (a) EGD for nickel oxalate dihydrate, (b) EGA by GC for anhydrous nickel formate (isothermal DSC; 480 K).

known mass of hydrate or decomposing compounds known to yield only one product e.g. CO_2 only or CO only.

Schmid and Felsche [26] used TG and DTA coupled with MS to study the thermal decompositions of hydrated copper(II) and cobalt(II) tartrates. Their results for the cobalt salt are illustrated in Fig. 10.7. The m/e traces of the strongest signals are shown and the initial endothermic process is clearly identified as dehydration only, while the higher temperature endothermic decomposition of the tartrate group gives rise to several fragments as shown. It is not possible to determine definitely whether these fragments are primary decomposition products or the result of secondary processes in the ion source of the MS. Detailed decomposition mechanisms were proposed [26].

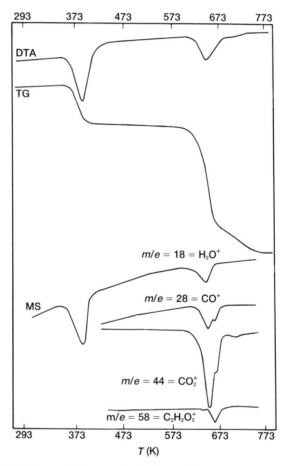

Figure 10.7 TG, DTA and MS curves for the thermal decomposition of hydrated cobalt tartrate, $Co(C_4H_4O_6)$. $2.5H_2O$, heated at 5 K min^{-1} in an argon atmosphere [26]. (With the permission of *Thermochimica Acta*, Elsevier, Amsterdam.)

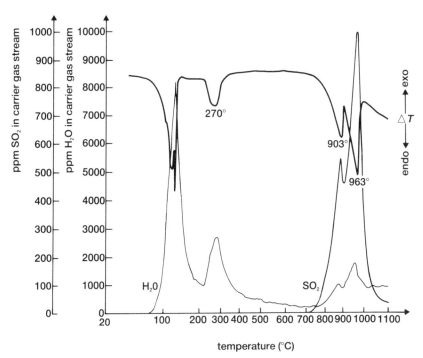

Figure 10.8 DTA–EGA of $CuSO_4 \cdot 5H_2O$ using non-dispersive infrared analyzers [14] for analysis of H_2O and SO_2. (With the permission of Wiley–Heyden Ltd.)

The DTA and EGA curves for $CuSO_4 \cdot 5H_2O$ using non-dispersive infrared analyzers [14] for H_2O and SO_2 are shown in Fig. 10.8. Once again the increased information obtainable by adding EGA to a TA system is illustrated.

Most of the references to the instrumental systems described above also give some illustrations of their application. The possibilities are virtually unlimited and in many situations (e.g. the degradation of polymers) it is essential, for practical applications, to have information on the gaseous products, especially hazardous products such as HCl and HCN.

10.5 References

1. Lodding, W. (ed.) (1967) *Gas Effluent Analysis*, Edward Arnold, London.
2. Dollimore, D., Gamlen, G. A. and Taylor, T. J. (1984) *Thermochim. Acta*, **75**, 59.
3. Holdiness, M. R. (1984) *Thermochim. Acta*, **75**, 361.
4. Emmerich, W. D. and Kaiserberger, E. (1979) *J. Therm. Anal.*, **17**, 197; (1982) *Proc. 7th ICTA*, Vol. 1, Wiley, Chichester, p. 279.
5. Barnes, P. A., Stephenson, G. and Warrington, S. B. (1981) *Proc. ESTA 2*, Heyden, London, p. 47.

6. Chan, K. C., Tse, R. S. and Wong, S. C. (1982) *Anal. Chem.*, **54**, 1238.
7. Clarke, E. (1981) *Thermochim. Acta*, **51**, 7.
8. Dyszel, S. M. (1983) *Thermochim. Acta*, **61**, 169.
9. Garn, P. D. and Anthony, G. D. (1967) *Anal. Chem.*, **39**, 1445.
10. Garn, P. D. (1964) *Talanta*, **11**, 1417.
11. Blandenet, G. (1969) *Chromatographia*, **2**, 184.
12. Hachenburg, H. (1973) *Industrial Gas Chromatographic Trace Analysis*, Heyden, London.
13. Thompson, B. (1977) *Fundamentals of Gas Analysis by Gas Chromatography*, Varian, Palo Alto.
14. Morgan, D. J. (1977) *J. Thermal Anal.*, **12**, 245.
15. Filsinger, D. H. and Bourrie, D. B. (1982) *Proc. 7th ICTA*, Vol. 1, Wiley, Chichester, p. 284.
16. Lephardt, J. O. (1982–3) *Appl. Spectrosc. Rev.*, **18**, 265.
17. Liebman, S. A., Ahlstrom, D. H. and Griffiths, P. R. (1976) *Appl. Spectrosc.*, **30**, 355.
18. Lephardt, J. O. and Fenner, R. A. (1980) *Appl. Spectrosc.*, **34**, 174.
19. Lephardt, J. O. and Fenner, R. A. (1981) *Appl. Spectrosc.*, **35**, 95.
20. Davidson, R. G. and Mathys, G. I. (1986) *Anal. Chem.*, **58**, 837.
21. Warrington, S. B. and Barnes, P. A. (1980) *Proc. 6th ICTA*, Vol. 1, Birkhaeuser Verlag, Basel, p. 327.
22. Gallagher, P. K., Gyorgy, E. M. and Jones, W. R. (1982) *J. Thermal Anal.*, **23**, 185.
23. Gallagher, P. K. and Gyorgy, E. M. (1980) *Proc. 6th ICTA*, Vol. 1, Birkhaeuser Verlag, Basel, p. 113.
24. Cote, R., Bale, C. W. and Gauthier, M. (1984) *J. Electrochem. Soc.*, **131**, 63.
25. Dessy, R. E. (1985) *Anal. Chem.*, **57**, 1188A, 1298A.
26. Schmid, R. L. and Felsche, J. (1982) *Thermochim. Acta*, **59**, 105.

Chapter 11

Less-common techniques

Under this heading, referring to Table 1.1, come emanation thermal analysis (ETA), thermomagnetometry, thermoelectrometry and thermosonimetry. These techniques are all gaining in popularity but usually require more specialized equipment which is often not commercially available as complete systems.

11.1 Emanation thermal analysis (ETA)

11.1.1 Introduction [1, 2]

Emanation thermal analysis (ETA) involves the measurement of the release of inert (and usually radioactive) gas from an initially solid sample, on heating. The rate of release of gas is used as an indication of the changes taking place in the sample. Most of the solids to be studied do not naturally contain inert gas. There are thus several techniques for incorporating inert gas.

11.1.2 Sample preparation

Techniques for incorporating inert gas in a solid sample may be divided broadly into two groups: (A) techniques for introducing the *parent* nuclide of the inert gas, e.g.

$$^{228}\text{Th} \xrightarrow{\ \alpha\ } \ ^{224}\text{Ra} \xrightarrow{\ \alpha\ } \ ^{220}\text{Rn}.$$

(B) techniques for introducing the inert gas itself. Labelling with parent nuclide gives a sample which is stable for longer as far as a source of inert gas is concerned. Gas is being formed continuously and will not all be lost in a single run on the sample. Under these headings, the following methods have been used.

A1 *Coprecipitation* of parent isotopes. The disintegration process ensures a random distribution of the products.

A2 *Impregnation* of the sample with a solution of parent isotope is used when coprecipitation is not possible. Parent atoms are distributed only on the surface. Disintegration results in penetration of the daughter atoms into crystallites, but unless the crystallites are small the distribution is non-uniform. The distribution may be altered by annealing the sample.

A3 *Nuclear reactions* may be used to produce the parent nuclide or the inert gas, e.g.

$$^{88}Sr \xrightarrow{\text{n},\alpha} {}^{85}Kr; \; {}^{44}Ca \xrightarrow{\text{n},\alpha} {}^{41}Ar$$

$$^{127}I \xrightarrow{\text{n,p}} {}^{128}I \longrightarrow {}^{128}Xe$$

B1 *Ion bombardment* is a common method of implanting inert gas atoms in solids. The ions may be generated in vacuum using an electrical discharge or a microwave plasma. The quantity of gas absorbed depends on the gas used, the energy of bombardment and the nature of the solid.

B2 *Diffusion* at high temperature and/or pressure is also sometimes possible. ^{85}Kr is usually used. (Solids labelled with ^{85}Kr are called kryptonates. $t_{1/2} = 10.76$ years for β-emission.)

B3 *Crystallization or sublimation* of the sample in an inert gas atmosphere may be more effective in incorporating the gas.

11.1.3 Measurement of gas release

Details of the gas release are very dependent upon the way in which the sample was originally labelled. Release may involve either bulk or defect diffusion. There may also be recoil ejection. The emanating power, E, is defined as

$$E = \dot{N}_{rel}/\dot{N}_{form}$$

where $\dot{N}_{rel}$ is the release rate and $\dot{N}_{form}$ is the formation rate. $\dot{N}_{form}$ may be measured by dissolution of the sample in acid or other solvent and measurement of the gas released.

ETA is usually carried out in conjunction with other TA techniques, e.g. Netzch markets ETA/DTA/EGA apparatus. Carrier gas, at an accurately controlled flow rate is used to carry released gas to suitable counting chambers. Rn, an α-emitter, requires a scintillation counter, or an ionization chamber or semi-conductor detector, while Geiger counters are used for Kr, Xe and Ar (β-emitters). For experiments lasting a long time compared with the half-life of the inert gas used, the decay of the measured gas should be taken into account.

11.1.4 Applications of ETA

In general, there is good agreement between ETA results and those of DTA and DTG. One of the major uses of ETA is in the characterization of powders.

Emanating power is related to surface area and hence changes in grain size and the occurrence of sintering during heating may be detected.

Figure 11.1 shows a series of isothermal ETA curves for NiO samples, prepared from the carbonate. These curves, obtained in a nitrogen atmosphere, were used [3] to determine the kinetic parameters for sintering of the samples. The apparent activation energy of 276 kJ mol^{-1} was reduced to 155 kJ mol^{-1} when a similar series of runs were done in oxygen. Phase changes also show up as changes in emanating power, e.g. the orthorhombic to rhombohedral phase transition of KNO_3 at 128°C (Table 4.2) is seen in Fig. 11.2.

ETA curves for decompositions (including dehydration) are similar to those obtained using DTA or DTG. Solid–gas reactions have also been studied, e.g. the oxidation of labelled metals or reduction of labelled oxides. ETA and EGA

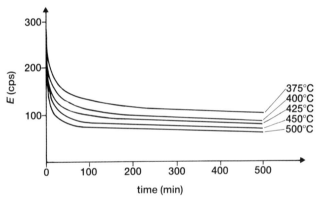

Figure 11.1 Isothermal ETA curves for NiO powders in nitrogen [2, 3]. (With the permission of Elsevier Scientific Publishers, Amsterdam.)

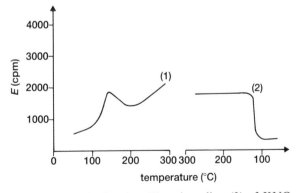

Figure 11.2 ETA traces for the heating (1) and cooling (2) of KNO_3 through the orthorhombic–rhombohedral phase transition at 128°C [2, 4]. (With the permission of Elsevier Scientific Publishers, Amsterdam.)

curves are in good agreement. Information on solid–liquid reactions, e.g. the hydration of cement, has been obtained. In both solid–gas and solid–liquid reactions the formation of a layer of solid product may hinder emanation.

For solid–solid reactions such as spinel formation [2, 5], e.g., $ZnO(s) + Fe_2O_3(s) \rightarrow ZnFe_2O_4(s)$, with ZnO labelled with ^{228}Th, the emanation of the individual components is checked first. ETA traces for the reaction are shown in Fig. 11.3. The oxides interact in a series of stages. Product begins to form by surface diffusion between 250 and 400°C as shown by the increased emanation (curve 3). The DTA and TD curves (curves 2 and 1 in Fig. 11.3) show no changes at these temperatures. Sharp changes in all of the curves in Fig. 11.3 are observed at 670 to 700°C. The exotherm on the DTA curve is small, but the

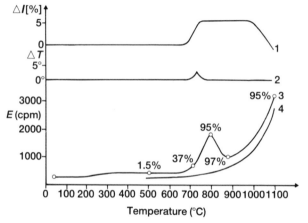

Figure 11.3 Spinel formation by the reaction of $ZnO–Fe_2O_3$ mixtures. Curve 1 is the TD, curve 2 the DTA and curve 3 the ETA trace, all at 10 K min^{-1}. The ETA trace during reheating of the reacted mixture is shown in curve 4 [2, 5]. (With the permission of Elsevier Scientific Publishers, Amsterdam.)

increased emanation (curve 3) shows up clearly. This corresponds to interaction by volume diffusion. The reaction is complete by about 800°C. The dilation shown in curve 1 is suggested [2] to be caused by the formation of a very finely powdered product which sinters at higher temperatures. The ETA curve during the second heating of the reaction mixture (curve 4) shows that reaction was complete during the first heating. The reactivity of different preparations of Fe_2O_3 has been compared using this technique [2].

The main drawback of ETA is that preparation and handling of samples requires all the usual radiochemical facilities and precautions. The amounts used in samples are so small that evolved gas, after dilution with carrier, does not form a hazard. Inert gases are not incorporated biologically and the decay products are stable, so hazards are reduced.

11.2 Thermomagnetometry (TM)

11.2.1 Introduction

When a sample is placed in a magnetic field, it may experience either attractive or repulsive forces. Attractive forces arise from three types of property of the sample: antiferromagnetism, paramagnetism and ferromagnetism (with ferromagnetism causing the strongest interaction). The diamagnetic properties of a sample, arising from the orbital motion of electrons, and hence present in all samples, but sometimes swamped by the other magnetic properties, give rise to weak repulsive forces. Antiferromagnetism occurs in only a few transition-metal compounds.

The volume magnetic susceptibility κ of a sample is defined as $\kappa = M/H$, where M is the magnetization and H is the magnetic field strength. The mass magnetic susceptibility χ is then $= \kappa/\rho$, where ρ is the density of the sample. Materials in which χ is positive are called paramagnetic, and those for which χ is negative are called diamagnetic. (An electronic microbalance is ideally suited for measurement of magnetic susceptibility [6].)

Different types of magnetic behaviour can be distinguished by their temperature dependence. For paramagnetic samples, the magnetic susceptibility decreases with absolute temperature according to the Curie–Weiss law: $\chi = C/(T - \theta)$, where C is the Curie constant and θ is the Weiss constant. The susceptibility of diamagnetic materials does not change much with temperature, while ferromagnetic materials have high susceptibilities at temperatures below what is known as the Curie point. Above their Curie points, ferromagnetic materials become paramagnetic and these magnetic transitions are used for temperature calibration of TG systems (chapter 3). It should also be noted that Curie points can be determined by impedance measurements [7].

11.2.2 Apparatus

Both the apparatus used, and the techniques and precautions required for thermomagnetometry are basically those required for thermogravimetry (TG) (chapter 3), with the addition of means of producing a strong magnetic field ($> 10^4$ gauss) around the sample. The apparent mass of the sample is then the sum of the actual sample mass and the magnetic force. Allowance obviously has to be made for the effect of the field on the sample container, which should thus have a low magnetic susceptibility e.g. quartz. The magnetic field may be applied periodically so that measurements of actual mass (TG) and apparent mass (TM) can be compared. Interference of the magnetic field with the balance mechanism is avoided by use of long suspension wires and suitable magnetic screening.

11.2.3 Applications of TM [12]

The major applications of TM have been to ferromagnetic samples where the magnetic effects are strongest. For example, TM has been used [8] to examine the spinel phases formed during the thermal decomposition of siderite ($FeCO_3$). The TM and TG curves shown in Fig. 11.4 for siderite samples heated in nitrogen, show the onset of decomposition at around 400°C. As the wustite (FeO) originally formed is oxidized by evolved CO_2, magnetite (Fe_3O_4) is formed. The apparent mass gain in the TM curve occurs as the magnetite nuclei grow and crystallize. When the temperature rises beyond the Curie point of magnetite, the TM trace coincides with the TG trace in the absence of a magnetic field. In oxygen, the wustite is oxidized so rapidly to haematite that the strongly magnetic spinel phase never has a chance to form.

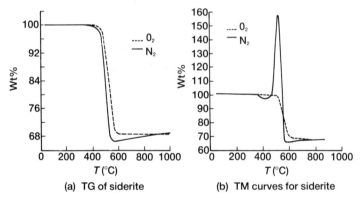

Figure 11.4 TM and TG curves of siderite ($FeCO_3$) heated in N_2 and in O_2 [8]. (With the permission of *Thermochimica Acta*, Elsevier, Amsterdam.)

The conditions for optimization of the formation of the ferrite $NiFe_2O_4$ from decomposition of the NiFe citrate have also been determined through use of TM [9].

TM provides a means of studying reactions in sealed systems [5], e.g. corrosion, provided that there is at least one ferromagnetic reactant or product. Figure 11.5 shows the apparatus used and the results obtained for the corrosion of carbon steel in an ammoniated EDTA solution.

Another application is in the determination of pyrite in coal [11], usually in combination with EGA. The residual ash from normal proximate analysis is heated in a reducing atmosphere (H_2/N_2) in the presence of a magnetic field (Fig. 11.6). From the apparent changes in mass, as the Fe_2O_3 is reduced to Fe and the product is cooled, the pyrites content can be calculated [11].

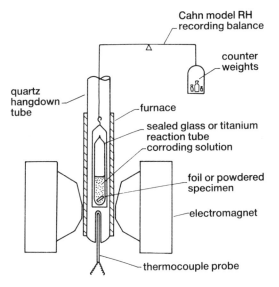

(a) TM apparatus as used for aqueous corrosion work

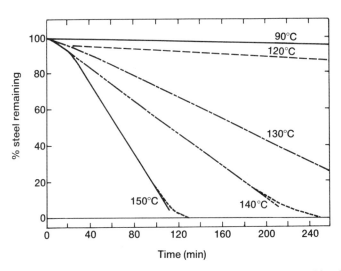

(b) Corrosion of 2 mil type 285 carbon steel in 5% ammoniated EDTA solutions

Figure 11.5 TM apparatus used, and results obtained for corrosion of carbon steel by ammoniated EDTA solution [10]. (With the permission of John Wiley & Sons, Chichester.)

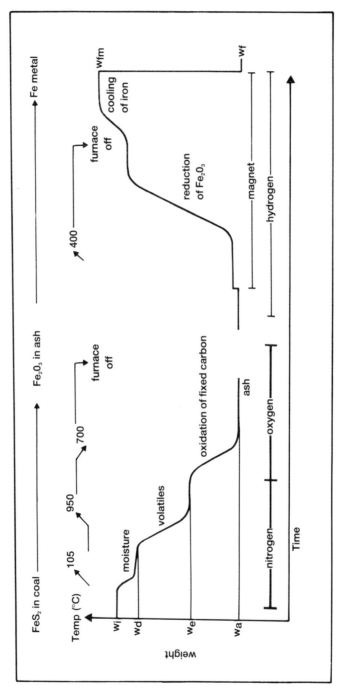

Figure 11.6 Determination of pyrites in coal using TM [11]. %FeS$_2$ = $((w_{fm} - w_c)/J_s)$ (molar mass FeS$_2$/molar mass Fe) (100/(w_i or w_d), J_s = saturation magnetization of iron = 218 emu g^{-1}. (With the permission of Wiley–Heyden Ltd.)

11.3 Thermoelectrometry

11.3.1 Introduction

The main electrical properties of the sample which may be measured as a function of temperature are resistance (or conductance) and capacitance. A further technique involving measurement of the EMF generated when two dissimilar metal electrodes are in contact with the sample during a heating programme, is known as thermovoltaic detection [13]. Most thermoelectrometry studies are carried out simultaneously with other techniques, especially DTA. TG studies on the decomposition of solids have also been carried out in the presence of applied electrical fields.

11.3.2 Apparatus

Both DTA and DSC cells have been modified for simultaneous measurements of electrical resistivity [14]. The system used [14] with the DuPont DSC is shown in Fig. 11.7. The DSC cell base is insulated from the rest of the cell with a thin glass slide. A thin piece of asbestos was used in the reference position to

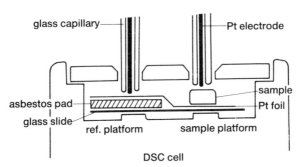

Figure 11.7 DuPont DSC cell modified for electrical resistivity measurements [14]. (With the permission of *Thermochimica Acta*, Elsevier, Amsterdam.)

compensate for the heat capacity of the sample. Platinum foil was used to form an electrical connection from the bottom of the sample to the top of the reference. Glass-insulated platinum wire electrodes contact the tops of the sample and the reference. The current through the sample was recorded as a function of temperature. The dimensions of the sample were then used to calculate the resistivity. Sample surfaces were coated with colloidal graphite to overcome contact resistance. The sensitivity of the DSC was reduced considerably by the presence of the glass slide.

Measurements of dielectric constant [15, 16] are based upon the measurement of the phase shift and attenuation of a sinusoidal AC signal after passage through the sample. The sample cell for concurrent dielectric measurement and

DTA [15] is illustrated in Fig. 11.8. A coaxial-type two-terminal electrode configuration is used. The inner electrode is a silver rod positioned symmetrically with respect to the outer electrode, which is a thin silver foil pressed against the walls of the cavity in the nickel block. The sample may be a liquid, a powder or a machined solid. Approximately 500 mg of sample is required. A separate smaller (50 mg) sample is used for concurrent DTA. The sample is connected to the inverting input of an operational amplifier, configured as a current-to-voltage converter with capacitive feedback. The measured phase-shift and attenuation of this network can be related [15] to the dielectric properties of the sample.

The apparatus can be operated in two modes: either using a continuous range of frequencies (50 Hz to 1 MHz) at selected temperatures, or over a continuous

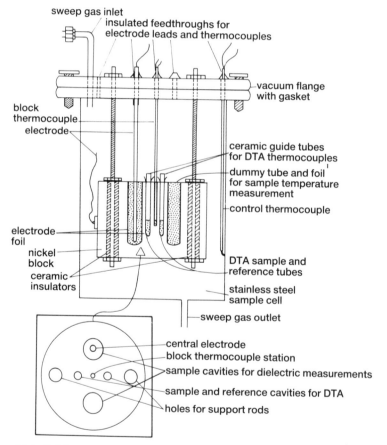

Figure 11.8 Cell for measurement of dielectric constant [15]. (With the permission of the American Chemical Society.)

range of temperatures at selected frequencies. Changes in the dielectric properties of the sample, brought about by phase transitions or chemical reactions are better resolved using the second mode. The instrument was calibrated with standard organic liquids.

In the earlier apparatus of Chiu [16] based on a DuPont 941 TMA module, an automatic digital capacitance bridge with three fixed frequencies (120, 400 and 1000 Hz), was used. Operation of the apparatus was checked using the transitions of KNO_3.

11.3.3 Applications of thermoelectrometry

Electrical conductivity measurements are useful in detecting the appearance of liquid phases in the reactions between initially solid reactants. It is also possible to monitor the loss of chemisorbed OH groups in quantities too small to be detected by TG. Changes in electrical conductivity during heating have also been attributed to changes in concentrations of crystal defects.

The dehydration of $CuSO_4.5H_2O$ has been studied by measuring the electrical resistance of a pressed sample simultaneously with thermodilatometry. The evolution of the five molecules of water in three steps is observed, as was found using TG (Fig. 3.9) and DSC (Fig. 4.16).

Measurements of resistivity against temperature for polymers [14] show sharp drops in resistivity at about the glass-transition temperature. The formation of conductive carbon chains by cross-linking and cyclization within the polymer can be detected by the gradual decrease of resistivity with increasing temperature. Such processes are not shown up by TG or DSC. A curve for PVC is shown in Fig. 11.9, with the glass transition and the region of dehydrochlorination marked. Thermoelectrometry is also useful in studying the effect of carbon black additives on the properties of polymers [14].

The thermal decomposition of oil shales has been shown [17], using thermoelectrometry, to be a two-step process involving the breakdown of an outer shell polar bridge (180–350°C) and cleavage of an inner naphthenic structure (350–500°C).

AC electrical conductivity measurements (up to 350°C) on the Group 1 metal and ammonium perchlorates [17] showed that identical charge conduction mechanisms attributed to movement of interstitial cations were present in all these salts.

Changes in the dielectric constant, ε', of a material with temperature can arise from changes in molecular orientation brought about by phase transitions or chemical reactions.

A curve of dielectric constant against temperature for $CuSO_4.5H_2O$ [18] has the same form as the TG curve (Fig. 3.9). The curve of $NaNO_2$ (120–200°C) shows a peak at $\sim 165°C$, of a form similar to that observed in the DTA curve, for the order–disorder crystal transition. This peak is attributed to Debye

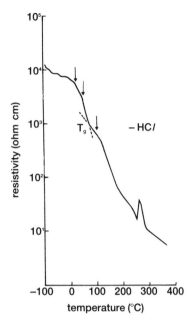

Figure 11.9 Thermoelectrometry curve for PVC [14]. (With the permission of *Thermochimica Acta*, Elsevier, Amsterdam.)

relaxation behaviour as N atoms diffuse along the *b*-axis in the crystal. Comparing the two curves, it was proposed that removal of H_2O from the copper sulphate crystals is a very fast diffusion process, so no dielectric relaxation is observed. Thus significant additional information is provided by the dielectric measurements.

Extension of the above studies [18] to include another known order–disorder ferroelectric crystal, $K_4Fe(CN)_6 . 3H_2O$, gave an endothermic DTA peak at $\sim 105°C$ with accompanying mass loss showing removal of $3H_2O$. Curves of dielectric constant against temperature show that prior to the dehydration, the H_2O molecules are involved in diffusion, similar to the order–disorder transition in $NaNO_2$, but escape of the H_2O leads to an ordered anhydrous compound.

11.4 Thermosonimetry (TS) and thermoacoustimetry

11.4.1 Introduction

In thermosonimetry (TS), sound waves emitted by a sample are measured as a function of temperature during a controlled heating programme. The sounds emitted arise from the release of thermal stresses in the sample and range in frequency from audio to several MHz. Stresses may be relieved by any of the thermal events described in Table 2.1, as well as by motion and creation of crystal defects. These last processes are not generally detectable by the more conventional thermal analysis techniques, on account of their low energy, and thus TS may be used, amongst other applications (section 11.4.4), to assess radiation damage, defect content and the degree of annealing of samples.

In thermoacoustimetry, the characteristics of imposed acoustic waves after passing through a sample are measured as a function of temperature while the sample is subjected to a controlled temperature programme. Further details of this technique are given briefly in section 11.4.3.

11.4.2 Apparatus for TS

The sounds are emitted as mechanical vibrations prior to and during thermal events in the sample. This sonic activity in the sample is picked up and transmitted by means of a specially adapted stethoscope. The mechanical waves are converted to electrical signals by conventional piezoelectric transducers. The output consists of a rapid cascade of decaying signals, which may be recorded in three ways: (i) as the mean rate (counts per second); (ii) as the mean amplitude level (energy); or (iii) by direct frequency monitoring.

The stethoscope is made of quartz (up to 1000°C) or ceramics or noble metals for higher temperatures. The sample is held in the sample head which acts as an acoustic transformer and is connected *via* a transmitting rod to a piezoelectric cell, fixed on a heavy recoil foundation and a seismic mount to prevent interference from external noise. A schematic diagram of the apparatus [19, 20] is shown in Fig. 11.10. The properties of the transducer may vary with temperature, so the waveguiding system is used to transmit the acoustic emissions from the heated sample to the transducer at ambient temperature. Contact surfaces must be well polished and thin films of silicone oil improve signal transfer. Care has to be taken that the switching of the temperature programmer does not register as noise on the transducer. Direct insertion of a thermocouple in the sample can cause severe mechanical damping, so the thermocouple is usually placed as close as possible to the sample without actually touching it. Alternatively a DTA head may be placed in the furnace for simultaneous TS-DTA measurements. Such a system has been used [20] to examine the transitions in the ICTA standards (Table 4.1).

Frequency distributions of the TS signals are obtained [20] by timing the intervals between the amplitude components of the decaying signal. The time intervals are converted into pulse heights which are fed to a multi-channel analyzer to give a display of the frequency distribution. Relationships are then sought between the frequency distributions and the processes occurring in the sample.

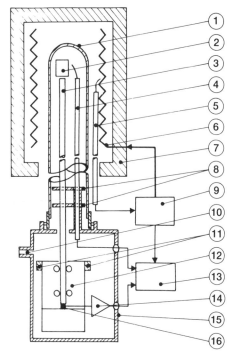

1. Protection tube of alumina
2. Sample
3. Stethoscope
4. Pt/10% Rh thermocouple
5. Furnace thermocouple
6. Heating element of silicon carbide
7. Furnace
8. Radiation shields
9. Temperature control system
10. Inlet for atmospheric control
11. Seismic mount of the pick-up system
12. Stethoscope mounting and cell basement
13. Amplifier system
14. Pre-amplifier
15. Vacuum sealed housing of the pick-up system
16. Piezoelectric cell

Different stethoscope sample holders

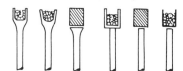

Figure 11.10 Apparatus for thermosonimetry [19]. (With the permission of John Wiley & Sons, Chichester.)

11.4.3 Apparatus for thermoacoustimetry

In the apparatus used by Mraz *et al.* [21] a pair of lithium niobate transducers are in contact (under a constant pressure of ~3 atm) with opposite faces of the sample (Fig. 11.11). One transducer induces the incident acoustic signal and the other detects the transmitted signal. Thermal expansion of the sample during the heating programme is monitored continuously with a linear variable differential transformer (LVDT), so that changes in sample dimensions can be allowed for in the calculations. Arrangements are made for atmosphere control around the sample.

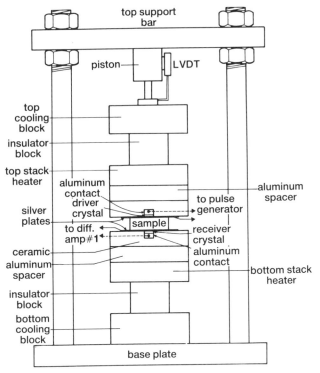

Figure 11.11 Apparatus for thermoacoustimetry [21]. (With the permission of *Thermochimica Acta*, Elsevier, Amsterdam.)

The incident signal is generated by a pulse generator. The transmitted signal received by the second transducer is inverted and amplified and an attenuated version of the driving pulse is added to this output signal. This summation procedure enables detection of the first-arrival times of both the compressional (P) and shear (S) waves. The velocities of the P and S waves and hence the various elastic moduli are then computed at set temperature intervals, and the final result is a plot of velocity or modulus against temperature. The instrument was calibrated with an aluminium standard for which the P and S wave velocities were accurately known.

11.4.4 Applications of TS and thermoacoustimetry

TS curves are usually used in combination with other thermoanalytical results to characterize a sample. TS curves on kaolins [19] show two regions of increased activity (Fig. 11.12). Comparing results obtained by TS with those obtained by TG and DTA, these regions have been identified as dehydroxylation (500–600°C) followed by recrystallization (980–985°C) to the metakaolin

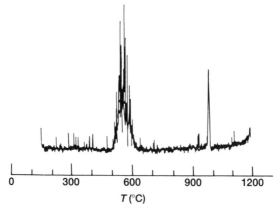

Figure 11.12 TS curve for a kaolin sample [19]. (With the permission of John Wiley & Sons, Chichester.)

structure. TS examination of the thermal decomposition of a powdered sample of brucite ($Mg(OH)_2$) [19], suggested that dehydroxylation occurs in a stepwise fashion with successive bursts of reaction as the sample breaks up. The barrier effect which has to be overcome was not identified.

The frequency distributions obtained [20] on heating samples of K_2SO_4 (a) and $KClO_4$ (b) to their transition temperatures of 582°C and 299°C, respectively, are shown in Fig. 11.13. The pre-transition activity of $KClO_4$ involves release of included fluid, while that in K_2SO_4 corresponds to generation of micro-cracks. The frequency distribution for K_2SO_4 shows fewer low frequency components. Frequency distributions for the dehydration steps in $CuSO_4 \cdot 5H_2O$ are qualitatively similar to those of $KClO_4$, again through the fluid loss processes.

Thermoacoustimetry has been used [21] to distinguish between grades of oil shales. Both the P and the S wave velocities decrease with increasing

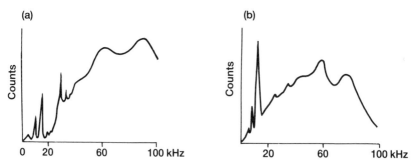

Figure 11.13 TS frequency distributions on heating. (a) K_2SO_4 and (b) $KClO_4$ [20]. (With the permission of John Wiley & Sons, Chichester.)

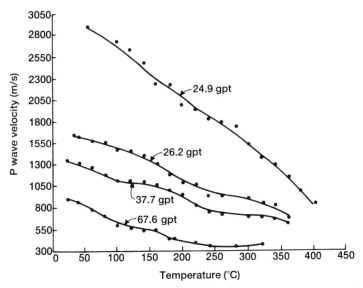

Figure 11.14 Thermoacoustimetry curves of oil shales [21]. (With the permission of *Thermochimica Acta*, Elsevier, Amsterdam.)

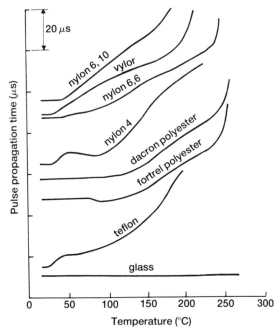

Figure 11.15 Thermoacoustimetry curves of synthetic fibres [22]. (With the permission of John Wiley & Sons, Chichester.)

temperature and with increasing organic content (Fig. 11.14). Results are very reproducible. Discontinuities and peaks in the plots are related to loss of water and decomposition of some hydrocarbon fractions.

Thermoacoustimetry has also been used [22], in combination with DTA, to examine the characteristics of synthetic fibres (Fig. 11.15). The increases in signal occur firstly at the glass-transition temperatures and then prior to melting. The glass fibre shows no changes in this temperature range.

11.5 References

1. Balek, V. (1978) *Thermochim. Acta*, **22**, 1; (1982) *Proc. 7th ICTA*, Vol. 1, Wiley, New York, p. 371; (1987) *Thermochim. Acta*, **110**, 222.
2. Balek, V. and Tolgyessy, J. (1984) *Emanation Thermal Analysis, Comprehensive Analytical Chemistry*, Vol. XIIC, Elsevier, Amsterdam.
3. Gourdier, F., Bussiere, P. and Imelik, B. (1967) *C. R. Acad. Sci.*, **264C**, 1624.
4. Balek, V. and Zaborenko, K. B. (1969) *Russian J. Inorg. Chem.*, **14**, 464.
5. Balek, V. (1970) *J. Am. Ceram. Soc.*, **53**, 540.
6. Daniels, T. (1973) *Thermal Analysis*, Kogan Page, London, p. 190.
7. Karmazin, E., Durand, B. and Romand, M. (1981) *Proc. 2nd ESTA*, p. 81.
8. Gallagher, P. K. and Warne, S. St. J. (1981) *Thermochim. Acta*, **43**, 253.
9. Tzehoval, H. and Steinberg, M. (1982) *Israel J. Chem.*, **22**, 227.
10. Charles, R. G. (1982) *Proc. 7th ICTA*, Vol. 1, p. 264.
11. Aylmer, D. and Rowe, M. W. (1982) *Proc. 7th ICTA*, Vol. 2, p. 1270.
12. Warne, S. St. J. and Gallagher, P. K. (1987) *Thermochim. Acta*, **110**, 269.
13. Wendlandt, W. W. (1984) *Thermochim. Acta*, **73**, 89; (1980) **37**, 117, 121.
14. Sircar, A. K., Lombard, T. G. and Wells, J. L. (1980) *Thermochim. Acta*, **37**, 315.
15. Rajeshwar, K. and co-workers, (1979) *Thermochim. Acta*, **33**, 157; (1978), **26**, 1; (1979) *Anal. Chem.*, **51**, 1149.
16. Chiu, J. (1974) *Thermochim. Acta*, **8**, 15.
17. Rajeshwar, K. and co-workers, (1980) *Nature* (London), **287**, 131; (1980) *J. Chem. Phys.*, **72**, 6678; (1980) *J. Phys. Chem. Solids*, **41**, 271; (1980) *Phys. Status Solidi*, **58**, 245.
18. Bristoti, A., Bonilla, I. R. and Andrade, P. R. (1976) *J. Thermal Anal.*, **9**, 93; (1975) **8**, 387.
19. Lonvik, K. and co-workers (1975) *Proc. 4th ICTA*, Vol. 3, p. 1089; (1980) *Proc. 6th ICTA*, Vol. 2, p. 313; (1982) *Proc. 7th ICTA*, Vol. 1, p. 306; (1982) *J. Therm. Anal.*, **25**, 109; (1987) *Thermochim. Acta*, **110**, 265.
20. Clark, G. M. (1981) *Proc. 2nd ESTA*, p. 85; (1978) *Thermochim. Acta*, **27**, 19.
21. Mraz, T., Rajeshwar, K. and Dubow, J. (1980) *Thermochim. Acta*, **38**, 211.
22. Chatterjee, P. K. (1975) *Proc. 4th ICTA*, Vol. 3, p. 835.

Chapter 12

The use of microcomputers in thermal analysis

12.1 Introduction

As with other fields, thermal analysis has been greatly influenced by the microcomputer revolution [1]. Wunderlich [2] has dealt with some of the more historical aspects of this impact. It is of interest that only just over ten years ago Wendlandt [3] reviewed the applications of digital and analogue computers in thermal analysis. Since that time analogue computers have faded from the scene and changes have been so rapid that equipment described in papers published even a few years ago can easily be obsolete. In dealing with obsolescence, one is, of course, referring to the on-line use of computers for direct data capture and interactive control of the instrument, as distinct from the more established off-line uses in data processing (Fig. 12.1).

Many of the manufacturers of thermal analysis equipment now offer both the hardware and the software necessary to carry out most thermal analyses with a high degree of automation, and to calculate the usual parameters from the captured data. As might be expected, these complete systems are not cheap, as prices include a component for software development, which is notoriously expensive. These systems are also specific to the particular manufacturer's equipment and, in some cases, the software may be difficult, or even impossible to modify for one's own requirements. Some calculations, such as the derivation of kinetic parameters under non-isothermal conditions [4], are areas of controversy and continuing development and it is necessary to know the approach and the algorithms being used in such software before any significance can be attached to the output.

For routine applications and/or in the absence of programming expertise, the computer systems provided by the manufacturers of the thermal analysis equipment are probably worth the additional expense. Some of these systems have been described in the literature, e.g. the DuPont 1090 system [5], the Mettler TA3000 system [6], and the Perkin-Elmer TADS system [7], but the latest information is best obtained from the manufacturers.

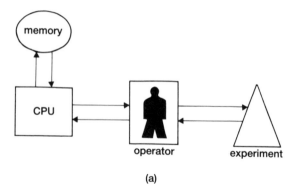

(a)

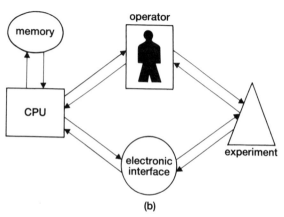

(b)

Figure 12.1 Use of computers [13]: (a) off-line, (b) on-line. (With the permission of the *Journal of Chemical Education.*)

For research purposes and/or when funds are not readily available but someone with an interest in computers is, an initial reduction in sophistication can probably be traded against increased flexibility, by coupling one of the many general-purpose microcomputers to whatever thermal analysis equipment is available. The prices of microcomputers have fallen to levels comparable with those for good chart recorders and their dedicated use should be easily justified in terms of the increased efficiency.

12.2 Hardware requirements [8, 9]

The hardware requirements for coupling of a computer to an instrument include the following aspects.

(1) Suitable analogue signals should be available from the instrument for analogue-to-digital conversion. Modern instruments may have special outputs in the usual ranges of 0 to 5 V, 0 to 10 V or −5 to +5 V for most analogue-to-digital converters (ADCs) (Fig. 12.2(a)). The output signals from older instruments were designed for potentiometric recorders in the mV range and will need high-gain, low-noise amplification. Interfaces containing variable gain amplifiers (which may be software controlled) in addition to ADCs are commercially available [8]. Certain instruments, usually with digital readouts, may have some outputs, e.g. temperature, in

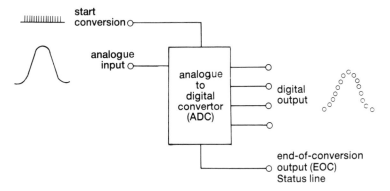

(a) Four-bit analogue-to-digital converter (ADC)

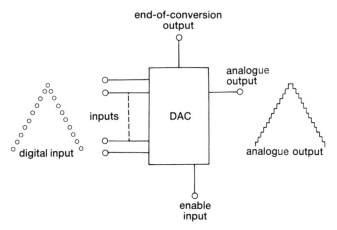

(b) Basic digital-to-analogue converter (DAC)

Figure 12.2 Computer interfacing [13]. (From Perone, S. P. and Jones, D. O. (1973) *Digital Computers in Scientific Instrumentation*, McGraw-Hill Book Co., New York, with permission.)

binary-coded decimal (BCD) form [9] to permit interfacing with other devices, including computers. Alternatively, the analogue signal may be converted to BCD through use of a suitable digital voltmeter (DVM). The interfacing of a BCD output to a computer is described by Blanck [10].

When only linear heating programmes are used in the thermal analysis, a direct reading of the temperature is not essential, provided that the time scale can be accurately related to the temperature (see (4) below).

(2) The resolution of ADCs [8] depends on the number of bits in the digital output e.g. an 8-bit ADC may convert the maximum input, 5 V say, into a number 255 (binary 1 1 1 1 1 1 1 1). Maximum resolution is then 1 in 256 or 19.5 mV. Variations of signal smaller than this value will thus not be detected. A 12-bit ADC gives a more acceptable maximum resolution of 1 in 4096, which is sufficient for most purposes, although 16-bit converters are available at much higher cost. For sequential capturing of several signals, an ADC with an adequate number of channels is required.

(3) In most thermal analyses there is no demand for rapid (<1 s interval) sampling of the signals, so a very fast (and hence more expensive) conversion time in the μs range is not necessary.

(4) Timing of sampling requires that the computer should have a real-time clock. In simpler systems, sampling may then be carried out by repeated reading of the clock until the set time is reached, at which time that ADC is read, after which the program returns to reading the clock until the next set time is reached. In more sophisticated systems an **interrupt** facility is available which signals the computer at preset intervals that sampling should take place. For the remainder of the interval the computer is free to carry out other operations and thus may do real-time processing of the data already collected, or may service the requirements of other instruments.

(5) For the control of instruments, for automation of analyses, the computer is required to do a calculation based on values of signals captured and to make a decision on the basis of the result (e.g. in a TG run, no further loss of mass may be occurring). The decision, in the form of a number, then has to be translated into some kind of control action through use of a digital-to-analogue converter (DAC) (Fig. 12.2(b)). The voltage output, corresponding to the number, may be used to operate a switch or valve, or to drive a stepping motor etc. For the DAC 8-bit resolution is usually ample.

Interactive control of the instrument during an analysis may be arranged by using one or more additional channels of the ADC to input signals, upon receipt of which action is taken. A channel of the DAC may be used to actuate an alarm, if necessary.

It is seldom necessary to do any of the design or building of interfacing equipment. Anything required can usually be quickly assembled from standard commercially-available modules [8].

12.3 Software requirements

When it comes to considering software requirements, it is very important, as James [1] puts it, 'not to reinvent wheels in computing'. Much software is available, admittedly of variable quality, but at prices that are low, through economics of scale, compared to the cost of development from scratch. The availability of a wide range of high-quality, general-purpose software should be one of the criteria in the original selection of the hardware.

Because of the generally slow rate of sampling required, the program for data acquisition can generally be written in a high level language such as BASIC. For faster sampling, ASSEMBLER or MACHINE CODE programs are usually necessary.

Again, because of the slow rate of sampling, some sort of averaging procedure can often be introduced at this stage to smooth the captured signal. Several rapid readings of the ADC output may be taken and the average stored.

12.4 Data storage

Data will generally be initially stored as an array within the random-access memory (RAM) of the computer (the size of which will limit the number of points that can be collected) and must then be stored more permanently, either before and/or after further processing. Storage may be on magnetic tape or, more efficiently, on disk. An alternative which has been suggested [11], is to link the dedicated microcomputer, *via* a communications interface, to a mainframe computer and to transfer the data to the larger computer to make use of its usually superior storage facilities, calculating ability and powerful software, as well as its peripherals, such as printer and plotter. This approach avoids the expense of having the dedicated microcomputer connected to its own storage device (although this is needed for so many purposes), printer and plotter.

During the storage process, all the relevant run parameters should be included for later reference. Back-up copies of important data are essential, especially when storage is on floppy disk. The ease and convenience of handling of data files may be a consideration in deciding upon the computer system to be purchased. (This is one of the weaknesses of Apple microcomputers.)

12.5 Data processing

The data processing required for thermal analysis in general, and especially for DSC and DTA, bears many resemblances to that developed for gas chromatography, e.g. the establishment of and correction of the signal for the baseline; detection of onset of peaks, peak maxima and the return to the baseline; resolution of overlapping peaks and determination of areas under peaks by numerical integration [12, 13]. The routines required for TG usually involve smoothing and numerical differentiation [14, 15], buoyancy corrections, and possibly polynomial regression. All of the above procedures are well documented (e.g. ref. 16).

Some programs in Applesoft BASIC are listed and described in Appendix B. Figure 12.3(a) shows a noisy DSC trace (obtained using program DSC) after correction for the rescanned baseline, using program OPERFILE. The trace was smoothed using one pass of program SMOOTHFILE (5-point Savitsky–Golay filter) (Fig. 12.3(b)) and five passes (Fig. 12.3(c)). Program BINARY FILTER was also used for smoothing of the original trace. The results for one, five and ten passes are shown in Figs 12.3(d), (e) and (f), respectively. The areas under the peaks were determined using program TRAPSUM, and are given on Fig. 12.3, to show the effect of smoothing on the measured area. Spikes on a trace are best removed manually or using program ROBUST FILTER. For more details see Appendix B. Other more specialized procedures, such as kinetic analysis under non-isothermal conditions, determination of purity by detailed analysis of the shape of melting endotherms, the determination of glass-transition temperatures and of heat capacities, may be developed to suit the user's needs. General procedures such as graphics display and plotting routines are obviously dependent upon the specific equipment in use.

The choice of programming language [17] will usually be some form of BASIC, because of the amount of software available, but a knowledge of FORTRAN is useful for implementation of published data processing routines [16], and more and more programs are being written in the more structured languages such as PASCAL, C, APL, ADA etc. [17]. Compilers are available for use with BASIC which give a considerable increase in speed.

12.6 Additional benefits

Once a microcomputer is available, even though much of its time may be dedicated to data capture, there will always be opportunities for use in additional (bonus) areas such as word processing, reference storage and possibly even searching of literature and data banks, as well as laboratory stock-control and computer-assisted instruction (CAI) [2].

12.7 Automation of thermal analysis equipment

Thermal analysis equipment does not generally lend itself to full automation in the sense of analysis of sample after sample without operator intervention [18].

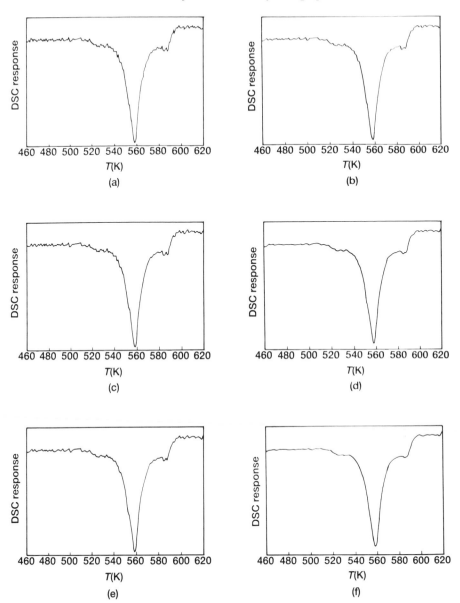

Figure 12.3 Digital smoothing of DSC traces and the effect of smoothing on the area (517.8 to 597.8 K) under the peak. (a) Original DSC trace (area = 100.0); (b) Savitsky–Golay 5-point filter 1 pass (area = 99.4); (c) Savitsky–Golay 5-point filter 5 passes (area = 98.4); (d) binary filter: 1 pass (area = 98.1); (e) binary filter: 5 passes (area = 96.0); (f) binary filter: 10 passes (area = 94.9).

Great advances are however being made in the programming of complex manipulations and developments can be expected in this direction. If there is some means of changing the sample automatically, there is usually no problem in arranging for automation of the heating/cooling programme and the switching of carrier gas as required. Most modern instruments allow for the storage, recall and implementation of such procedures. A multi-specimen 'carousel' thermobalance has been designed by Ferguson *et al.* [19] (Fig. 12.4), which allows twenty specimens, mounted on arms radiating from a central

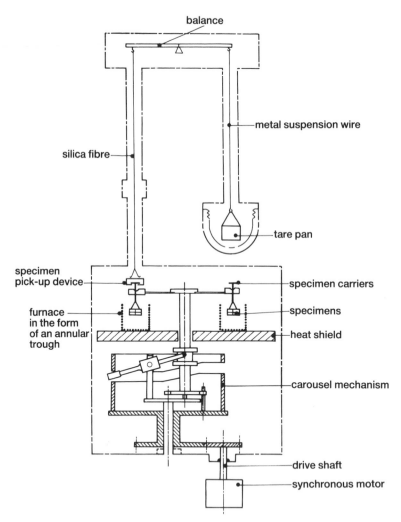

Figure 12.4 Schematic diagram of multi-specimen carousel thermobalance [19]. (With the permission of John Wiley & Sons, Chichester.)

shaft, to be weighed sequentially. There is obviously a limit to the sorts of treatment which the samples can receive between weighings, but they can all be exposed to a similar environment.

The Perkin-Elmer Corporation has recently announced a robotic system which enables 48 samples to be run consecutively on their DSC without operator intervention. The system, which is shown in Fig. 12.5, has a removable 48-position sample carousel and a pneumatically controlled sampling arm. The arm automatically selects any desired sample from the carousel, places it in the DSC sample holder and closes the cover. Control is *via* the data station's microcomputer, so that once the sample is loaded, it is subjected to programmed treatment as desired. On completion of this treatment, the sample is removed from the DSC and replaced in the carousel. The sequence in which the samples are handled and their individual treatments can be programmed and the data obtained are automatically stored on disk. The system is obviously at its most efficient when used for quality control with one procedure applied to many similar samples.

Figure 12.5 Perkin-Elmer DSC-4 Robotic System. (With the permission of Perkin-Elmer Corporation.)

12.8 References

1. James, E. B. (1982) *Chem. Br.*, **18**, 620.
2. Wunderlich, B. (1982) *Int. Lab.*, October, 32.
3. Wendlandt, W. W. (1973) *Thermochim. Acta*, **5**, 225.
4. Blazejowski, J. (1984) *Thermochim. Acta.*, **76**, 359.
5. Burroughs, P. and Leckenby, J. N. (1981) *Proc. 2nd Eur. Symp, on Thermal Analysis (ESTA)*, p. 67.
6. Fruh, P. and Widmann, G. (1982) *Int. Lab.*, Jan/Feb, 48.
7. Brennan, W. P., Fyans, R. L. and Mayer, J. S. (1981) *Proc. 2nd Eur. Symp. on Thermal Analysis (ESTA)*, p. 76; (1982) *Proc. 7th ICTA*, p. 255.
8. Bogdam, M. R. and Warme, P. K. (1983) *CAL*, **1**, 40.
9. Malmstadt, H. V., Enke, C. G. and Crouch, S. R. (1981) *Electronics and Instrumentation for Scientists*, Benjamin/Cummings, Menlo Park.
10. Blanck, H. F. (1984) *J. Chem. Educ.*, **61**, 533.
11. Tseng, H. P., Christman, P. G. and Edgar, T. F. (1981) *Chem., Biomed., and Environ. Instrumentation*, **11**, 377.
12. Cooper, J. W. (1977) *The Minicomputer in the Laboratory*, Wiley, New York.
13. Perone, S. P. and Jones, D. O. (1973) *Digital Computers in Scientific Instrumentation*, McGraw-Hill, New York.
14. Savitsky, A. and Golay, M. J. E. (1964) *Anal. Chem.*, **36**, 628.
15. Bussian, B.-M. and Hardle, W. (1984) *Appl. Spectrosc.*, **38**, 309.
16. Norris, A. C. (1981) *Computational Chemistry*, Wiley, Chichester.
17. Owen, G. S. (1984) *J. Chem. Educ.*, **61**, 440.
18. Millett, E. J. (1976) *J. Phys. E., Sci. Instr.*, **9**, 794.
19. Ferguson, J. M., Livesey, P. M. and Mortimer, D. (1972) *Prog. Vacuum Microbalance Techniques*, Vol. 1, (eds. T. Gast and E. Robens) Heyden, London, p. 87.

Chapter 13

Reaction kinetics from thermal analysis

13.1 Introduction

The rate of a general homogeneous reaction of the form

$$A \rightarrow B + C$$

is conventionally measured by following the decrease in concentration of reactant A or the increase in concentration of either product B or C at constant temperature. A rate equation of the form

Rate $= k\,f$(concentrations of reactants and products)   (T constant)

is then determined from experiment. The rate coefficient, k, is a function of temperature:

$$k = A\,e^{-E/RT} \quad (\text{or } k = AT^m e^{-E/RT})$$

and by carrying out a series of experiments over a range of different but constant temperatures, the Arrhenius parameters, E the activation energy and A the pre-exponential or frequency factor, can be determined.

In thermal analysis the reactions studied are almost invariably heterogeneous reactions and the reaction temperature is usually being continuously increased or decreased according to some set (usually linear) programme. In spite of these differences, much effort has been directed at obtaining kinetic information from the results of thermal analysis experiments. Numerous papers have appeared and are still appearing on this topic and it is a field of considerable controversy. What follows is only an introductory and fairly selective account of some of the problems and the approaches suggested for their solution. More detailed accounts can be found in the references quoted.

13.2 Heterogeneous reactions

When a solid sample is heated, one of the many possible changes which it may undergo is decomposition (Table 2.1). Information on the kinetics and mechanisms of solid decompositions is of both practical and theoretical importance [1]. For the heterogeneous reaction

$$A(s) \rightarrow B(s) + C(g)$$

$$\text{e.g.,} \quad CaCO_3(s) \rightarrow CaO(s) + CO_2(g)$$

the concept of concentration no longer has the same significance and the progress of reaction has to be measured in some other way. Usually the fractional reaction, α, is defined in terms of the change in mass of the sample ($\alpha = (m_0 - m)/(m_0 - m_f)$ where m_0 is the initial mass and m_f the mass of the sample when reaction is complete) or equivalent definitions in terms of amounts of gas evolved or heat absorbed or evolved. The heterogeneity of such reactions is not restricted to the participation of the different phases represented above. There are also the variations in properties with direction (anisotropy) in the crystal structure of a single pure solid, as well as the presence of numerous impurities and structural defects, including surfaces, edges, dislocations and point defects, in any real solid sample. Such factors have no analogy in homogeneous reactions, but have marked effects on the physical properties of solids, especially their thermal stability [2].

Numerous observations confirm that decomposition of solid reactants generally is initiated at defective regions of the crystal such as the surface or, more specifically, points of emergence of dislocations at the surface. Nuclei of solid product, B, are thus formed, the gaseous product escapes (sometimes with difficulty) and the resulting disruption causes strain in the neighbouring regions of unreacted A, resulting in growth of the nuclei (Fig. 13.1). The shape of these nuclei will be governed by the crystal structure, in that decomposition in some directions will occur more readily than in others. The detailed geometry of the processes of nucleation and growth [3, 4] leads to specific predictions of the rate at which the product gas is evolved. In addition to considering the geometry of the reactant/product interface, account has to be taken of the chemical reactions taking place at or near the interface [4] and physical processes such as diffusion and heat transfer [5].

13.3 Formulation of the problem

A kinetic study thus involves measurement of α either as a function of time, t, at constant temperature, or as a function of temperature, T, which is increased according to some heating programme (usually linear), $\phi = dT/dt$. The isothermal method, α *vs. t*, corresponds to the conventional concentration *vs. t* curve of homogeneous kinetics, while the dynamic method, i.e. measurement of α *vs. T*, is the basis of thermal analysis (Figs 13.2(a) and (b)).

Kinetic analysis of both isothermal and dynamic results involves attempting to relate the experimentally observed α, t or α, T values with values predicted for

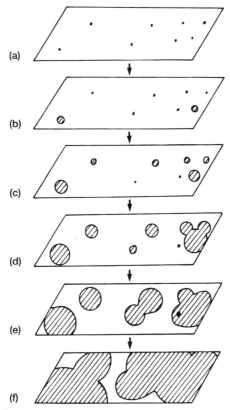

Figure 13.1 Formation and growth of nuclei of product in the decomposition of solids: (a) nucleation sites, (b) first nuclei formed, (c) growth and further nucleation, (d) overlap of nuclei, (e) ingestion of a nucleation site, (f) continued growth.

a limited set of models [1] based on processes of nucleation and growth, diffusion, or some simpler geometrical forms of progress of the reactant/product interface. The expressions derived from these models can all be written in their integral forms (at constant T)

$$f(\alpha) = k(t - t_0)$$

or differential forms

$$d\alpha/dt = k\, g(\alpha)$$

(see Table 13.1 and Fig. 13.3).

The effect of temperature is introduced through use of the Arrhenius equation, $k = A e^{-E/RT}$ (or sometimes $k = A T^m e^{-E/RT}$ [6–8]) so that

$$d\alpha/dt = A e^{-E/RT} g(\alpha)$$

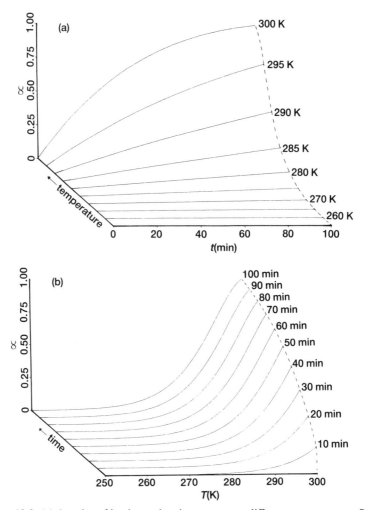

Figure 13.2 (a) A series of isothermal α-time curves at different temperatures [56]. (b) The constant time contours derived from (a). (With the permission of the *Journal of Chemical Education*.)

The validity of applying the Arrhenius equation to heterogeneous reactions has been questioned [5, 9], but the parameters E and A do have practical value even if their theoretical interpretation is difficult.

For reversible reactions, the rate of reaction will depend on the partial pressure of the gaseous product. The rate equation should thus include allowance for this, in terms of some function $h(p)$

$$\mathrm{d}\alpha/\mathrm{d}t = A\mathrm{e}^{-E/RT}\, g(\alpha)\, h(p)$$

Table 13.1 Broad classification of solid-state rate expressions

		$f(\alpha) = kt$	$g(\alpha) = \dfrac{1}{k}\dfrac{d\alpha}{dt}$
1. Acceleratory α–time curves			
P1	power law	$\alpha^{1/n}$	$n(\alpha)^{(n-1)/n}$
E1	exponential law	$\ln \alpha$	α
2. Sigmoid α–time curves			
A2	Avrami–Erofe'ev	$[-\ln(1-\alpha)]^{1/2}$	$2(1-\alpha)(-\ln(1-\alpha))^{1/2}$
A3	Avrami–Erofe'ev	$[-\ln(1-\alpha)]^{1/3}$	$3(1-\alpha)(-\ln(1-\alpha))^{2/3}$
A4	Avrami–Erofe'ev	$[-\ln(1-\alpha)]^{1/4}$	$4(1-\alpha)(-\ln(1-\alpha))^{3/4}$
B1	Prout–Tompkins	$\ln[\alpha/(1-\alpha)]$	$\alpha(1-\alpha)$
3. Deceleratory α–time curves			
3.1 based on geometrical models			
R2	contracting area	$1-(1-\alpha)^{1/2}$	$2(1-\alpha)^{1/2}$
R3	contracting volume	$1-(1-\alpha)^{1/3}$	$3(1-\alpha)^{2/3}$
3.2 based on diffusion mechanisms			
D1	one-dimensional	α^2	$1/2\alpha$
D2	two-dimensional	$(1-\alpha)\ln(1-\alpha)+\alpha$	$(-\ln(1-\alpha))^{-1}$
D3	three-dimensional	$[1-(1-\alpha)^{1/3}]^2$	$\frac{3}{2}(1-\alpha)^{2/3}(1-(1-\alpha)^{1/3})^{-1}$
D4	Ginstling–Brounshtein	$(1-2\alpha/3)-(1-\alpha)^{2/3}$	$\frac{3}{2}(1-\alpha)^{-1/3}-1)^{-1}$
3.3 based on 'order of reaction'			
F1	first order	$-\ln(1-\alpha)$	$1-\alpha$
F2	second order	$1/(1-\alpha)$	$(1-\alpha)^2$
F3	third order	$[1/(1-\alpha)]^2$	$(1-\alpha)^3$

This complication is usually ignored and this may be justifiable when working in vacuum or with a strong flow of inert gas through the sample.

For dynamic measurements, the usual approach is to write

$$d\alpha/dT = (d\alpha/dt)(dt/dT) = (1/\phi)(d\alpha/dt)$$

where $\phi = dT/dt$ is the heating rate. The heating rate is usually maintained constant, although other programmes, e.g. hyperbolic with $1/T = u - vt$ where u and v are constants, have been considered [10, 11] and Rouquerol *et al.* [12] have proposed heating the sample in such a way that reaction takes place at constant rate.

The relationship between $d\alpha/dT$ and $d\alpha/dt$ is one of the areas of controversy [13–17] but if the above procedure is accepted, and most treatments proceed on that basis, then

$$d\alpha/dT = (1/\phi)(d\alpha/dt) = (A/\phi)\,e^{-E/RT}g(\alpha) \tag{13.1}$$

Separating variables

$$d\alpha/g(\alpha) = (A/\phi)\,e^{-E/RT}\,dT$$

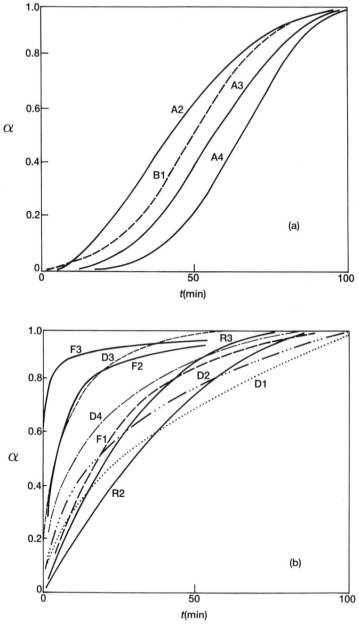

Figure 13.3 α-time plots for the model expressions (Table 13.2): (a) sigmoid group, (b) deceleratory group (based on α=0.98 at t=100).

Integrating between the limits, $\alpha = 0$ at $T = T_0$ and $\alpha = \alpha$ at $T = T$

$$\int_0^\alpha (1/g(\alpha))\, d\alpha = \int_{T_0}^T (A/\phi)\, e^{-E/RT} dT$$

$$f(\alpha) = \int_{T_0}^T (A/\phi)\, e^{-E/RT}\, dT = \int_0^T (A/\phi)\, e^{-E/RT}\, dT \qquad (13.2)$$

$$\left(\text{since } \int_0^{T_0} (A/\phi)\, e^{-E/RT}\, dT = 0 \right)$$

The unknowns are thus A, E and the form of either $f(\alpha)$ or $g(\alpha)$. (Note that there is no standardization on this terminology and in some papers $f(\alpha)$ and $g(\alpha)$ are used in exactly the opposite sense to that given here.) Often, following homogeneous kinetics, $g(\alpha)$ is taken to be $(1-\alpha)^n$, so that n, the apparent 'order of reaction', becomes the third unknown. A more general function, suggested by Sestak and Berggren [18] is

$$g(\alpha) = \alpha^m (1-\alpha)^n (-\ln(1-\alpha))^p$$

which increases the number of unknowns, but allows for all the usual models of solid state reactions. The unknowns then have to be determined from experimental measurements which can be converted to values of α and/or $d\alpha/dt$ at temperatures T, obtained at a set heating rate ϕ. The methods used in the analysis of the non-isothermal kinetic data are then usually divided into differential methods, based on use of equation (13.1), or integral methods, based on use of equation (13.2).

A brief history of the development of non-isothermal kinetic analysis has been given [19] and contributions to the field continue to appear in the current literature. The International Confederation for Thermal Analysis (ICTA) has appointed a committee to investigate and report on the field. In this introductory account only the main approaches will be outlined. More detail can be found in various reviews [1, 17, 19–21]. Some attention will be given to those approaches which have become 'named' methods and are often quoted and sometimes incorporated in software packages (chapter 12). Such methods will be amongst those to be fully scrutinized by the above-mentioned ICTA committee [58]. Their mention here implies no more than that they have been widely used. Objections that have been raised to the use of these specific methods will be mentioned where possible. In addition, the more general objections to the non-isothermal approach should be kept in mind. Some of these objections have been summarized by Garn [5] and concern the problems of measurement of sample temperature when allowance has to be made for heat transfer from the furnace to the outer regions of the sample and then into the sample; the self-cooling or self-heating of the sample during reaction, and the removal of evolved product gases from the vicinity of the sample and the influence of these products on the rate of reaction when the reaction has a high degree of reversibility.

Attempts have been made to determine all three unknowns, A, E and the form of $f(\alpha)$ or $g(\alpha)$, from a single dynamic (i.e. α, T) run. This is only possible if assumptions are made about the form of $f(\alpha)$ or $g(\alpha)$. It has often been assumed, without much justification, that reaction is analogous to a first-order process ($g(\alpha) = 1 - \alpha$ and $f(\alpha) = -\ln(1-\alpha)$).

If one is prepared to relax the constraint that data from a single run only are to be used, considerable simplification is possible. The three possibilities for two runs are that they should be: (i) both isothermal – which corresponds to conventional kinetic studies and, although it is probably the most reliable method, will not be discussed further here; (ii) one dynamic and one isothermal run; and (iii) two dynamic runs at different heating rates. Once more than one run is permitted, the approach is usually to eliminate the unknown form of the model $f(\alpha)$ or $g(\alpha)$ by comparing measurements made at a common value of α under the two sets of different conditions.

In summary then, the approaches may be classified as either differential or integral methods and, within each group, as either single-run or multiple-run methods.

13.4 Differential methods

These are methods based on equation (13.1)

$$d\alpha/dT = (A/\phi)\, e^{-E/RT}\, g(\alpha) \qquad (13.1)$$

A direct approach [22] has been to measure a series of α, T and $(d\alpha/dT)$ values at a given heating rate ϕ (with the practical problems of integration or differentiation and their associated errors, depending upon the initial form of the experimental data). If a form of $g(\alpha)$ is assumed, then from equation (13.1)

$$y = \ln[(d\alpha/dT)/g(\alpha)] = \ln(A/\phi) - (E/R)(1/T)$$

so that a plot of y against $(1/T)$ leads to values of E and A.

Piloyan *et al.* [23] suggested the simplification that for values of $\alpha < 0.5$, $g(\alpha)$ could be taken as approximately constant and E and A could be determined from a plot of $\ln(d\alpha/dT)$ against $1/T$.

An extension of this approach, when more than one set of data is available, is the method of Carroll and Manche [24] in which $d\alpha/dT$ is measured at a fixed value of α from a series of thermal analyses at different heating rates, ϕ. Then $\ln(\phi\, d\alpha/dT)$ is plotted against $1/T$ and E is calculated from the slope.

The often-quoted Freeman–Carroll method [25] assumes $g(\alpha) = (1 - \alpha)^n$ and considers incremental differences in $(d\alpha/dT)$, $(1 - \alpha)$ and $(1/T)$ and leads to the expression

$$\Delta \ln(d\alpha/dT) = n \Delta \ln(1-\alpha) - (E/R) \Delta(1/T)$$

This expression can be used to determine the value of E by plotting either $[\Delta \ln(d\alpha/dT)/\Delta \ln(1-\alpha)]$ against $[\Delta (1/T)/\Delta \ln(1-\alpha)]$ or $[\Delta \ln(d\alpha/dT)/\Delta(1/T)]$ against $[\Delta \ln(1-\alpha)/\Delta(1/T)]$. Sestak *et al.* [19] have discussed some improvements, including extension to other models. It is worth noting that for a constant heating rate ϕ, $\Delta \ln(d\alpha/dT) = \Delta \ln(d\alpha/dt)$.

Criado *et al.* [26] have demonstrated that the Freeman and Carroll treatment does not allow an nth order kinetic equation to be distinguished successfully from other kinetic models. This was confirmed by Jerez [27] who pointed out the large errors involved in the regression procedure, and suggested a modification in the calculation of E and n involving the use of the point at which the rate is a maximum and the centre of gravity of the cluster of experimental points.

Van Dooren and Müller [28] have examined the effects of sample mass, particle size and heating rate on the determination of n, E and A from DSC data using the Freeman and Carroll procedure. Plots using points from the complete DSC peaks were curved, so different sections of the DSC peak were considered separately. Different values of the kinetic parameters were obtained for the different section of the peak and these values also varied with sample mass, particle size and heating rate.

Several approaches involve use of the second derivative of equation (13.1), or the version with $g(\alpha) = (1-\alpha)^n$, with respect to temperature [19, 20] or with respect to α [20], in spite of the problems of obtaining accurate values of second derivatives. Using

$$d\alpha/dT = (A/\phi)\, e^{-E/RT}\, (1-\alpha)^n \tag{13.3}$$

gives

$$\frac{d^2\alpha}{dT^2} = \frac{d\alpha}{dT}\left(\frac{E}{RT^2} - \left(\frac{d\alpha}{dT}\right)\frac{n}{(1-\alpha)}\right)$$

and since this derivative must be zero at the inflexion point of a TG curve or the maximum of a DSC peak

$$E/(RT_{max}^2) = (d\alpha/dT)_{max}\, (n/(1-\alpha_{max})) \tag{13.4}$$

from which E may be calculated if n is known and T_{max}, $(d\alpha/dT)_{max}$ and α_{max} are measured [29]. Combining equations (13.3) and (13.4) gives

$$(A/\phi)\, e^{-E/R} T_{max}\, n(1-\alpha_{max})^{n-1} = E(RT_{max}^2)$$

and since $(1-\alpha_{max})$ is a constant for a given value of n, the Kissinger [30] method of obtaining a value for E is to plot $\ln(\phi/T_{max}^2)$ against $1/T_{max}$ for a series of runs at different heating rates, ϕ. The slope of such a plot is $-E/R$.

Augis and Bennett [31] have modified the Kissinger treatment for use with the Avrami–Erofe'ev model, which applies in so many solid-state reactions.

They plot $\ln(\phi/(T_{max} - T_0))$ against $1/T_{max}$ where T_0 is the initial temperature at the start of the heating programme, instead of $\ln(\phi/T_{max}^2)$ against $1/T_{max}$.

Elder [32] has generalized the Kissinger treatment to make it applicable to the full range of kinetic models. The generalized equation is

$$\ln(\phi/T_{max}^{m+2}) = \ln(AR/E) + \ln(L - E/RT_{max})$$

where m is the temperature exponent of the pre-exponential term in the modified Arrhenius equation and is often taken as zero, and

$$L = -g(\alpha_{max})/(1 + mRT_{max}/E)$$

This correction term was found to be relatively small, but helps in distinguishing between similar models. The values of E obtained were not very sensitive to incorrect choice of model.

Although originally derived as an integral method, the Ozawa treatment [33] is also applicable to derivative curves and is similar to the Kissinger method. In the Ozawa method $\ln \phi$ is plotted against $1/T_{max}$ and the slope of this plot is again $-E/R$.

Van Dooren and Müller [34] studied the effects of sample mass and particle size on the determination of kinetic parameters from DSC runs using the methods of Kissinger and of Ozawa. It was found that both sample mass and particle size could influence the values of the kinetic parameters, but the extent of these effects varied from one substance examined to another. The two methods gave similar values for E with slightly lower precision for the Kissinger method. It was suggested that temperatures at $\alpha = 0.5$ (half conversion) should be used in place of T_{max}.

The Borchardt and Daniels method [35, 36], originally developed for DTA studies of homogeneous liquid-phase reactions, involves calculating the rate coefficient, k, from the expression [37]

$$k = [(KSV/m_0)^{n-1} (C \, d\Delta T/dt + K \, \Delta T)]/[(K(S-s) - C \, \Delta T)]^n$$

where V is the volume of the sample and m_0 the initial amount of sample in moles, S is the total DTA peak area, s is the partial area up to the time t at which the peak height T and the slope $d\Delta T/dt$ are measured, K is the heat transfer coefficient and C is the overall heat capacity of the sample and the holder. Because of the problems of obtaining values of K and C, the assumption that $n = 1$ and the approximation [36] that $C(d\Delta T/dt)$ and $C\Delta T$ are small compared to the terms to which they are added, or from which they are subtracted, are made, so that

$$k = \Delta T/(S - s)$$

For DSC the equivalent form would be

$$k = (dH/dt)/(S - s)$$

The values of k obtained are then used in a conventional Arrhenius plot. Shishkin [38] has discussed the similarities between the Borchardt and Daniels and the Kissinger approaches.

Another method based on a second derivative is that of Flynn and Wall [39]. They differentiated equation (13.3), rearranged as

$$T^2(\mathrm{d}\alpha/\mathrm{d}T)=(AT^2/\phi)\,\mathrm{e}^{-E/RT}\,(1-\alpha)^n$$

with respect to α, to give

$$\frac{\mathrm{d}}{\mathrm{d}\alpha}\left(T^2\,\frac{\mathrm{d}\alpha}{\mathrm{d}T}\right)=(E/R)+2T+\left(\frac{n}{(1-\alpha)}\right)\left(\frac{\mathrm{d}\alpha}{\mathrm{d}(1/T)}\right)$$

At the beginning of decomposition, i.e. low α, the last term on the right-hand side is negligible and even $2T\ll E/R$. They then took finite differences instead of the derivative, so that

$$\Delta(T^2\,\mathrm{d}\alpha/\mathrm{d}T)/\Delta\alpha=(E/R)+2\bar{T}$$

where $\bar{T}$ was the average temperature over the interval. The value of E was then calculated from the slope of a plot of $\Delta(T^2\mathrm{d}\alpha/\mathrm{d}T)$ against $\Delta\alpha$ for the beginning of the reaction only, or, if one starts from $\alpha=0$ where $\mathrm{d}\alpha/\mathrm{d}T=0$, $T^2(\mathrm{d}\alpha/\mathrm{d}T)$ may be plotted against α.

Some of the more common differential methods are summarized in Table 13.2.

13.5 Integral methods

Use of equation (13.2)

$$f(\alpha)=\int_0^\alpha (1/g(\alpha))\,\mathrm{d}\alpha=(A/\phi)\int_{T_0}^T \mathrm{e}^{-E/RT}\,\mathrm{d}T \tag{13.2}$$

obviously involves evaluation of the temperature integral. Usually the limits of integration are changed to $\int_0^T \mathrm{e}^{-E/RT}\,\mathrm{d}T$ if α is zero up to T_0. The problem is simplified by introducing the variable $x=E/RT$ so that

$$\int_0^T \mathrm{e}^{-E/RT}\,\mathrm{d}T=(E/R)\int_x^\infty (\mathrm{e}^{-x}/x^2)\,\mathrm{d}x=(E/R)p(x).$$

Then equation (13.2) becomes

$$f(\alpha)=(AE/R\phi)p(x) \tag{13.5}$$

Tables of values of the integral $p(x)$ have been provided [40, 41]. A lot of attention has been directed at finding suitable approximations for the above temperature integrals [1, 19, 20, 42, 43]. Gorbachev [44] has suggested that

Table 13.2 A selection of differential methods of non-isothermal kinetic analysis (linear plots: $y = mx + c$)

Name	y	x	Slope $= m$	Intercept $= c$	Ref.
Achar, Brindley, and Sharp	$\ln[(d\alpha/dT)/g(\alpha)]$	$1/T$	$-E/R$	$\ln(A/\phi)$	22
Piloyan	$\ln(d\alpha/dT)$ $(\alpha < 0.5)$	$1/T$	$-E/R$	$\ln(A/\phi) + \ln(g(\alpha))$	23
Carroll and Manche	$\ln(\phi\, d\alpha/dT)$ (at fixed α)	$1/T$	$-E/R$	$\ln(A) + \ln(g(\alpha))$	24
Freeman and Carroll	$[\Delta\ln(d\alpha/dT)/\Delta\ln(1-\alpha)]$	$[\Delta(1/T)/\Delta\ln(1-\alpha)]$	$-E/R$	n	25
	or $[\Delta\ln(d\alpha/dT)/\Delta(1/T)]$	$[\Delta\ln(1-\alpha)/\Delta(1/T)]$	n	$-E/R$	
Fuoss, Sayler, and Wilson	$(d\alpha/dT)_{max}(n/(1-\alpha_{max}))$	$(1/T_{max})^2$	E/R	0	29
Kissinger	$\ln(\phi/(T_{max})^2)$	$1/T_{max}$	$-E/R$	0	30
Ozawa	$\ln\phi$	$1/T_{max}$	$-E/R$	$\ln(AE/R) + \ln(g(\alpha_{max}))$	33
Borchardt and Daniels	$\ln(\Delta T/(S-s))$	$1/T$	$-E/R$	$\ln A$	35
	or $\ln((dH/dt)/(S-s))$	$1/T$	$-E/R$	$\ln A$	36
Flynn and Wall	$\Delta(T^2\, d\alpha/dT)$	$\Delta\alpha$	$(E/R)+2T$	0	39

there is little value in trying to find more accurate approximations considering the nature of the original α, T data.

Zsakó [20] has suggested sub-classification of integral methods on the basis of the means of evaluation of the temperature integral. The three main approaches are: (i) use of numerical values of $p(x)$; (ii) use of series approximations for $p(x)$; (iii) use of approximations to obtain an expression which can be integrated. The series most commonly used for approximating $p(x)$ are [20] the asymptotic expansion

$$p(x) = (e^{-x}x^2)[1 - (2!/x) + (3!/x^2) - (4!/x^3) + \cdots + (-1)^n((n+1)!/x^n) + \cdots]$$

or, with better results [45], the Schlömlich expansion

$$p(x) = (e^{-x}/x(x+1))[1 - (1/(x+2)) + (2/(x+2)(x+3))$$
$$- (4/(x+2)(x+3)(x+4)) + \cdots]$$

Taking only the first two terms of each series

$$p(x) \approx (x-2)\,e^{-x}/x^3$$

or

$$p(x) \approx e^{-x}/(x(x+2))$$

Other approximations are the empirical formula [46]

$$p(x) \approx e^{-x}/((x-d)(x-2))$$

with $d = 16/(x^2 - 4x + 84)$, or [47]

$$p(x) \approx e^{-x}/(x(x^2+4x)^{1/2})$$

Doyle's approximation [48] (for $x > 20$) is

$$\log_{10} p(x) \approx -2.315 - 0.4567x$$

Coats and Redfern [49] suggested an asymptotic expansion of the temperature integral, giving

$$f(\alpha) = (ART^2/\phi E)[1 - (2RT/E)]\,e^{-E/RT}$$

which, assuming $2RT/E \ll 1$, leads to

$$\ln f(\alpha) = \ln(AR/\phi E) + 2\ln T - (E/RT)$$

or

$$\ln(f(\alpha)/T^2) = \ln(AR/\phi E) - (E/RT)$$

One approach for arriving at an integrable expression [20, 43] is to use an expansion for $1/T$ in equation (13.2), with the aim of obtaining a simple function of T, $\theta(T)$ in equation (13.2) rewritten as [43]

$$\ln f(\alpha) = \ln C_0 + C_1\,\theta(T)$$

where C_0 is a constant containing A and C_1 a constant containing E.

Van Krevelen *et al.* [50] used

$$(1/T) = (1/T_r)[1 - (T - T_r)/T_r + \cdots]$$

while Horowitz and Metzger [51] used

$$(1/T) = (1/T_r)[1 - \ln(T/T_r) + \cdots]$$

where T_r is an arbitrarily defined reference temperature somewhere within the extremes of temperature covered by the experiment [43]. Van Krevelen's treatment leads to $C_1 = E/RT_r + 1$ and $\theta(T) = \ln T$, while Horowitz and Metzger's treatment gives $C_1 = E/R(T_r)^2$ and $\theta(T) = T$. The problem with these approaches is the arbitrariness of the reference temperature. Use has been made of the temperature at which the reaction rate is a maximum. Kassman [43, 52] has shown that equation (13.2) can be made integrable by using $\theta(T) = 1/T$. This leads to the integrated form

$$f(\alpha) = [Ae^2(T_r)^2/(\phi((E/R) + 2T_r))]e^{-((E/R) + 2T_r)/T}$$

hence

$$\ln f(\alpha) = -((E/R) + 2T_r)/T + \text{constant.}$$

Kassman also showed that the reference temperature can be optimized in the above expression, or those of Van Krevelen or Horowitz and Metzger above, by calculating the geometric mean of the temperature interval

$$T_{gm} = (T_{lowest} \, T_{highest})^{1/2}$$

and then using

$$T_r = T_{gm} [1 - (T_{gm}/C_1) + 2(T_{gm}/C_1)^2]$$

where C_1 is the observed slope of the fitted line when $\ln f(\alpha)$ is plotted against $1/T$. Kassman [43] claims that this method is superior to the others in determining values of E and A. He also examined the problems of discriminating amongst the models, $f(\alpha)$, and showed that it was not possible to distinguish models within the various groups in Table 13.1 from a single non-isothermal data set.

Zsakó [20, 53] suggested the following practical procedure starting from equation (13.5) in the form

$$\ln(f(\alpha)) - \ln(p(x)) = \ln(AE/R\phi) = B$$

where B is a constant for a given reaction and constant heating rate, ϕ. Various functions of α were assumed and combined with selected values of E. For each experimental α value, measured at temperature T, one thus obtains a value of $f_1(\alpha)$ and hence $\ln(f_1(\alpha))$. Combination of the corresponding experimental T value with an assumed E value $= E_1$, provides a value of $x_1 = E_1/RT$ and hence $p(x_1)$ and $\ln(p(x_1))$. If the choice of both the model, $f_1(\alpha)$ and the activation energy, E_1, was correct

$$\ln(f_1(\alpha)) - \ln(p(x_1)) \text{ should be a constant, } = \ln(AE_1/R\phi),$$

from which A can be calculated. Otherwise a new choice has to be made. Even using computers, such a search is time consuming.

Satava and Skvara [54] developed a graphical method to simplify the search. Tables of $\log(f(\alpha))$ for the various models are available [40], or can be readily generated, together with tables of $\log(p(x))$ against T for fixed values of E [19, 40]. The experimental α values are thus converted to $\log(f(\alpha))$ values for various functions of α, and are then plotted against the experimental T values, on the same scale as plots of $-\log(p(x))$ against T for various values of E (Fig. 13.4). One of these sets of plots should be on transparent paper. With the T scales matched, the plots are moved relative to each other in the y-direction until one of the $\log(f(\alpha))$ against T curves matches one of the $-\log(p(x))$ against T curves.

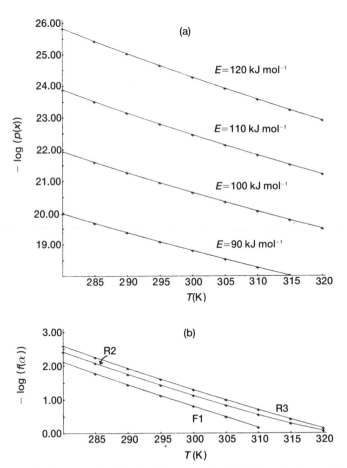

Figure 13.4 The plots required for the Satava method [3]. One plot is usually done on transparent paper so that they can be superimposed for matching [56]. (With the permission of the *Journal of Chemical Education*.)

The displacement between the y-scales is then equal to B, from

$$\log(f(\alpha)) - \log(p(x)) = B$$

Hence the value of A can be calculated. The drawbacks are: (i) the difficulties in distinguishing between models [4], and (ii) the use of discrete E values.

Relaxation of the constraints to allow data from one dynamic and one isothermal run, introduces considerable simplification. Under isothermal conditions $(T = T_0)$, $\alpha = \alpha_i$ at $t = t_i$ (Fig. 13.5) and $f(\alpha_i) = A\,e^{-E/RT_0}t_i$.

For the dynamic run at heating rate ϕ, $\alpha = \alpha_i$ at $T = T_i$, and $f(\alpha_i) = (AE/R\phi)\,p(x_i)$ where $x_i = E/RT_i$. Hence

$$A\,e^{-E/RT_0}t_i = (AE/R\phi)\,p(x_i)$$

or

$$t_i = (E/R\phi)e^{E/RT_0}p(x_i)$$

and

$$\log t_i = \log(E/R\phi) + (E/2.303\,RT_0) + \log(p(x_i))$$

Evaluation of $p(x_i)$ requires an estimate of E, but using the Doyle [40]

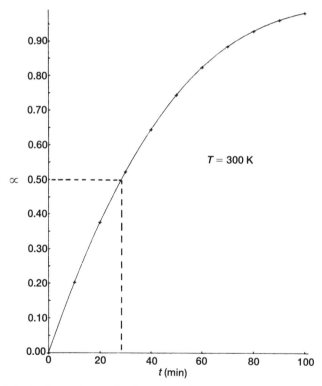

Figure 13.5 An isothermal curve for the contracting volume expression, R3 [56]. (With the permission of the *Journal of Chemical Education*.)

approximation that $\log(p(x)) = -2.315 - 0.4567x$ (if $x > 20$ and the temperature interval scanned in the dynamic runs is restricted to < 100 K)

$$\log t_i = \log(E/R\phi) + (E/2.303RT_0) - 2.315 - (0.4567E/R)(1/T_i) = m(1/T_i) + c$$

A plot of $\log t_i$ against $(1/T_i)$ should thus be linear (Fig. 13.6) with slope $m = -0.4567E/R$, from which a value of E can be calculated. Thus at the cost of an additional run it is possible to eliminate assumptions about the form of $f(\alpha)$.

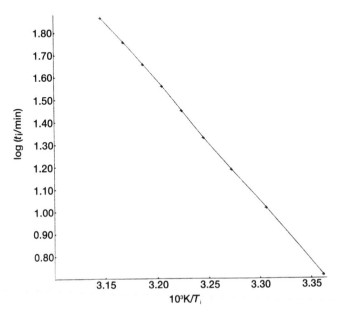

Figure 13.6 Doyle's method [5] of plotting $\log t_i$ against $1/T_i$ (see text) [56]. (With the permission of the *Journal of Chemical Education*.)

If instead of using data from one dynamic and one isothermal run, we use two dynamic runs at different heating rates ϕ_1 and ϕ_2, then, following Gyulai and Greenhow [55], equation (13.2) is written as

$$f(\alpha) = (A/\phi) \int_{T_0}^{T} e^{-E/RT} \, dT = (A/\phi) \int_{0}^{T} e^{-E/RT} \, dT = (A/\phi)b$$

where

$$b = \int_{0}^{T} e^{-E/RT} \, dT \quad \text{and} \quad \int_{0}^{T_0} e^{-E/RT} \, dT \approx 0$$

The method is again based on comparison of runs at common values of $\alpha = \alpha_i$. At heating rate ϕ_1, $\alpha = \alpha_i$ at $T = T_1$ but at heating rate ϕ_2, $\alpha = \alpha_i$ at $T = T_2$. Hence $f(\alpha_i) = (A/\phi_1)b_1 = (A/\phi_2)b_2$. A whole series of b_1/b_2 ratios is possible as the value of α_i has not been specified, i.e. $(\phi_1/\phi_2)=(b_1/b_2)_i$. Tables of $-\log b$ for various values of T and E are given [55]. Pairs of corresponding temperatures $(T_1$ and $T_2)_i$ at common values of α_i are determined from the two dynamic runs (Fig. 13.7). For some value of $E=E_j$ say, then combination with $(T_1$ and $T_2)_i$ results in values $(b_1$ and $b_2)_{ij}$. $\log(b_2/b_1)_{ij}$ is then plotted against E_j (Fig. 13.8) and the unknown value of E is determined from the point at which $\log(\phi_2/\phi_1)=\log(b_2/b_1)_{ij}$.

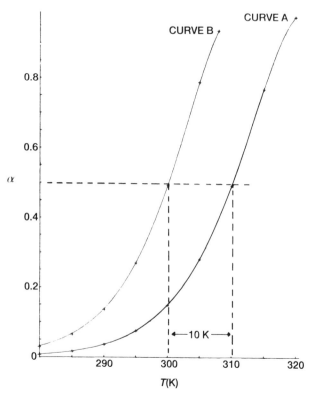

Figure 13.7 Non-isothermal curves for the contracting volume expression, R3 [56]. Heating rates: curve A 1.000 K min^{-1}, curve B 0.257 K min^{-1}. (With the permission of the *Journal of Chemical Education.*)

13.6 The influences of various parameters on the shapes of theoretical thermal analysis curves

With there being so many parameters in equations (13.1) or (13.2), it is useful to see what effect varying one of these parameters at a time has on the shape of a TG or DSC curve [19, 20, 53]. The most fundamental variable is the form of the

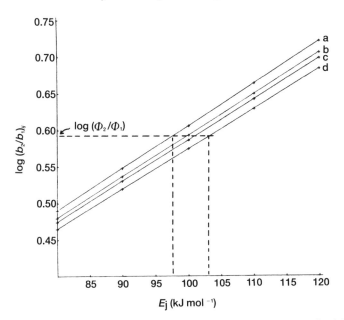

Figure 13.8 Gyulai and Greenhow's method [6]. Curves vary slightly (a–d) with value of α chosen [56]. (With the permission of the *Journal of Chemical Education*.)

mechanistic model $f(\alpha)$ or $g(\alpha)$. The shapes of the theoretical isothermal α–time curves for various models [1, 56] are shown in Fig. 13.3. It is not always an easy task to distinguish amongst the models, even under isothermal conditions [57]. The shapes of these curves are, of course, considerably altered under non-isothermal conditions and theoretical TG curves [20, 32] for various models are given in Figs 13.9 and 13.10. The models based on apparent order of reaction, n, (even if fractional) i.e. F1, F2, R2 and R3, Fig. 13.9, are difficult to distinguish at low values of α. Distinguishability improves for higher orders at higher values of α. The diffusion models, D2, D3 and D4, give generally lower onset temperatures and flatter curves (Fig. 13.10) than the nth-order group, while the Avrami–Erofe'ev models have higher onset temperatures and steeper curves. Figure 13.11 shows the differential curves corresponding to the integral curves given in Fig. 13.10.

The next stage is to choose one model and use this to examine the influence of the other variables, the heating rate ϕ, the activation energy, E, and the pre-exponential factor, A. Zsakó [20] has done this for the first-order (F1) model. Although this is not a very realistic model it is often assumed to apply as a first approximation. Figure 13.12 shows the regular effect on the theoretical first-order TG curve, of doubling the heating rate in the range $\phi = 1$ to 16 K min^{-1}. Decreasing the pre-exponential factor by factors of 10 in the range $A = 10^{16}$ to

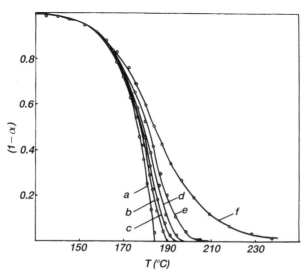

Figure 13.9 Influence of reaction order, n, on the shape of theoretical TG curves [20]. $E=142$ kJ mol^{-1}, $A=10^{14}$ s^{-1}, $\phi=4$ K min^{-1}. Order n: $a=0$; $b=1/3$; $c=1/2$; $d=2/3$; $e=1$; $f=2$. (From Zivkovic, Z. D. (Ed.) (1984), *Thermal Analysis*, University of Beograd, Yugoslavia, with permission.)

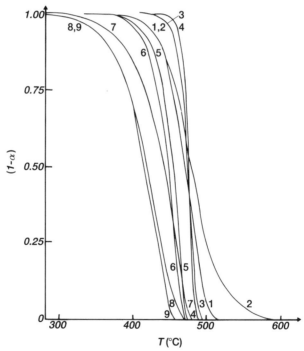

Figure 13.10 Integral curves for various models [32]. $\phi=10$ K min^{-1}, $E=220$ kJ mol^{-1}, $A=10^{15}$ min^{-1}. 1 to 9: models F1, A2, A3, R2, R3, D2, D3 and D4, respectively. (With the permission of Wiley–Heyden Ltd.)

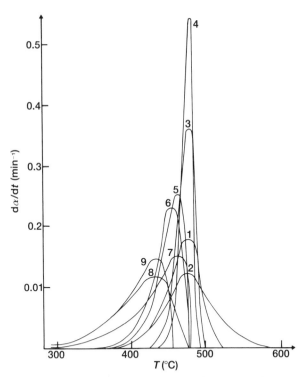

Figure 13.11 Differential curves for various models [32]. $\phi = 10$ K min^{-1}, $E = 220$ kJ mol^{-1}, $A = 10^{15}$ min^{-1}. 1 to 9: models F1, A2, A3, R2, R3, D2, D3 and D4, respectively. (With the permission of Wiley–Heyden Ltd.)

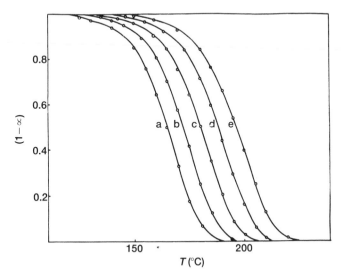

Figure 13.12 Influence of heating rate, ϕ, on the shape of theoretical TG curves [20]. $n = 1$, $E = 142$ kJ mol^{-1}, $A = 10^{14}$ s^{-1}. ϕ(K min^{-1}): $a = 1$; $b = 2$; $c = 4$; $d = 8$; $e = 16$. (From Zivkovic, Z. D. (Ed.) (1984) *Thermal Analysis*, University of Beograd, Yugoslavia, with permission.)

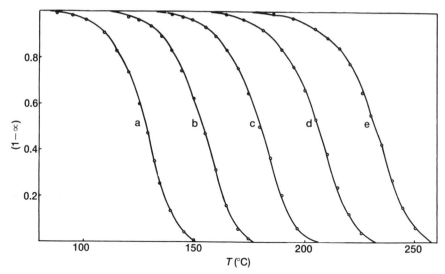

Figure 13.13 Influence of pre-exponential factor, A, on the shape of theoretical TG curves [20]. $n=1$, $E=142$ kJ mol^{-1}, $\phi=4$ K min^{-1}. $A(\text{s}^{-1})$: $a=10^{16}$; $b=10^{15}$; $c=10^{14}$; $d=10^{13}$; $e=10^{12}$. (From Zivkovic, Z. D. (Ed.) (1984) *Thermal Analysis*, University of Beograd, Yugoslavia, with permission.)

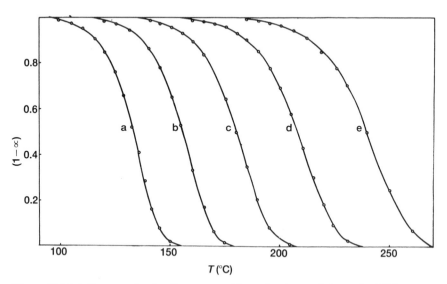

Figure 13.14 Influence of activation energy, E, on the shape of theoretical TG curves [20]. $n=1$, $A=10^{14}$ s^{-1}, $\phi=4$ K min^{-1}. E (kJ mol^{-1}): $a=126$; $b=134$; $c=142$; $d=151$; $e=159$. (From Zivkovic, Z. D. (Ed.) (1984) *Thermal Analysis*, University of Beograd, Yugoslavia, with permission.)

10^{12} s^{-1} also affects mainly the onset temperature and the acceleratory portion of the curves (Fig. 13.13) with the remaining segments being almost parallel. Very similar behaviour is observed for curves with the activation energy increasing in steps of 8.3 kJ mol^{-1} (2 kcal mol^{-1}) as shown in Fig. 13.14.

The overall shape of the thermal analysis curve is thus determined by the mechanistic model applying, while the position of this curve on the temperature axis is determined by the values of E, A and, to a lesser extent, the heating rate ϕ.

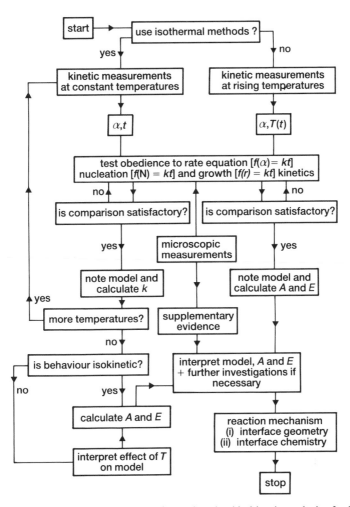

Figure 13.15 Schematic representation of steps involved in kinetic analysis of solid state reactions [3]. (With the permission of John Wiley & Sons, Chichester.)

13.7 Conclusion

Galwey [3] has given a flow diagram (Fig. 13.15) summarizing the steps involved in the kinetic analysis of solid-state reactions. Only non-isothermal methods have been discussed here, and the methods have been selected to illustrate the approaches that may be used. Numerous variations and modifications have been and are being published. More information will be found in refs 1 and 7–9. Garn [5] has provided a comprehensive and critical review of the many factors which must be taken into account when assessing the reliability of kinetic parameters derived from dynamic measurements. Two reasons given [5] for possible failure of a method of kinetic analysis are: (i) the approximations used may not be valid, and (ii) the model used in deriving the mathematics may not take into account physically real factors of the experiment such as the heat transfer and atmosphere control problems.

13.8 References

1. Brown, M. E., Dollimore, D. and Galwey, A. K. (1980) *Reactions in the Solid State, Comprehensive Chemical Kinetics*, Vol. 22, Elsevier, Amsterdam.
2. Boldyrev, V. V. (1986) *Thermochim. Acta*, **100**, 315.
3. Galwey, A. K. (1982) *Proc. 7th Int. Conf. Thermal Analysis*, Wiley-Heyden, Chichester, Vol. 1, p. 38.
4. Galwey, A. K. (1985) *Thermochim. Acta*, **96**, 259.
5. Garn, P. D. (1972) *Crit. Rev. Anal. Chem.*, **3**, 65.
6. Dollimore, D., Gamlen, G. A. and Taylor, T. J. (1982) *Thermochim. Acta*, **54**, 181.
7. Petty, H. R., Arakawa, E. T. and Baird, J. K. (1977) *J. Thermal Anal.*, **11**, 417.
8. House, J. E. Jr., (1981) *Thermochim. Acta*, **48**, 165.
9. Arnold, M., Veress, G. E., Paulik, J. and Paulik, F. (1982) *Thermochim. Acta*, **52**, 67; (1981) *Anal. Chim. Acta*, **124**, 341.
10. Varhegyi, G. (1982) *Thermochim. Acta*, **57**, 247.
11. Popescu, C. and Segal, E. (1982) *J. Thermal Anal.*, **24**, 309.
12. Reading, M., Dollimore, D., Rouquerol, J. and Rouquerol, F. (1984) *J. Thermal Anal.*, **29**, 775.
13. Gyulai, G. and Greenhow, E. J. (1974) *Talanta*, **21**, 131.
14. MacCullum, J. R. (1982) *Thermochim. Acta*, **53**, 375.
15. Garn, P. D. (1974) *J. Thermal Anal.*, **6**, 237.
16. Sestak, J. and Kratochvil, J. (1973) *J. Thermal Anal.*, **5**, 193.
17. Sestak, J. (1984) *Thermophysical Properties of Solids, Comprehensive Analytical Chemistry*, Vol. XIID, Elsevier, Amsterdam.
18. Sestak, J. and Berggren, G. (1971) *Thermochim. Acta*, **3**, 1.
19. Sestak, J., Satava, V. and Wendlandt, W. W. (1973) *Thermochim. Acta*, **7**, 447.
20. Zsakó, J. (1984) *Thermal Analysis*, (ed. Z. D. Zivkovic) University of Beograd, Bor, Yugoslavia, p. 167.
21. Blazejowski, J. (1981) *Thermochim. Acta*, **48**, 109; (1984) **76**, 359.
22. Achar, B. N., Brindley, G. W. and Sharp, J. H. (1966) *Proc. Int. Clay Conf. Jerusalem*, **1**, 67; Sharp, J. H. and Wentworth, I. A. (1969) *Anal. Chem.*, **41**, 2060.

23. Piloyan, F. O., Ryabchikov, I. O. and Novikova, O. S. (1966) *Nature* (London), **212,** 1229.
24. Carroll, B. and Manche, E. P. (1972) *Thermochim. Acta*, **3,** 449.
25. Freeman, E. S. and Carroll, B. (1958) *J. Phys. Chem.*, **62,** 394; (1969) **73,** 751.
26. Criado, J. M., Dollimore, D. and Heal, G. R. (1982) *Thermochim. Acta*, **54,** 159.
27. Jerez, A. (1983) *J. Thermal Anal.*, **26,** 315.
28. Van Dooren, A. and Müller, B. W. (1983) *Thermochim. Acta*, **65,** 269.
29. Fuoss, R. M., Sayler, O. and Wilson, H. S. (1964), *J. Polym. Sci.*, **2,** 3147.
30. Kissinger, H. E. (1956) *J. Res. Nat. Bur. Stand.*, **57,** 217; (1957) *Anal. Chem.*, **29,** 1702.
31. Augis, J. A. and Bennett, J. E. (1978) *J. Thermal Anal.*, **13,** 283.
32. Elder, J. P. (1985) *J. Thermal Anal.*, **30,** 657; (1984) *Analytical Calorimetry*, Vol. 5, (eds. P. S. Gill and J. F. Johnson) Plenum, New York, p. 269.
33. Ozawa, T. (1965) *Bull. Chem. Soc. Jpn.*, **38,** 1881; (1970) *J. Thermal Anal.*, **2,** 301.
34. Van Dooren, A. A. and Müller, B. W. (1983) *Thermochim. Acta*, **65,** 257.
35. Borchardt, H. J. and Daniels, F. (1957) *J. Am. Chem. Soc.*, **79,** 41.
36. Reed, R. L., Weber, L. and Gottfried, B. S. (1965) *Ind. Eng. Chem. Fundamentals*, **4,** 38.
37. Sharp, J. H. (1972) *Differential Thermal Analysis*, Vol. 2, (ed. R. C. Mackenzie) Academic Press, London, p. 47.
38. Shishkin, Y. L. (1985) *J. Thermal Anal.*, **30,** 557.
39. Flynn, J. H. and Wall, L. A. (1966) *J. Res. Nat. Bur. Stand.*, **70A,** 487.
40. Doyle, C. D. (1961) *J. Appl. Polym. Sci.*, **5,** 285.
41. Zsakó, J. (1968) *J. Phys. Chem.*, **72,** 2406.
42. Blazejowski, J. (1981) *Thermochim. Acta*, **48,** 125.
43. Kassman, A. J. (1985) *Thermochim, Acta*, **84,** 89.
44. Gorbachev, V. M. (1982) *J. Thermal Anal.*, **25,** 603.
45. Doyle, C. D. (1965) *Nature* (London), **207,** 290.
46. Zsakó, J. (1975) *J. Thermal Anal.*, **8,** 593.
47. Balarin, M. (1977) *J. Thermal Anal.*, **12,** 169.
48. Doyle, C. D. (1962) *J. Appl. Polym. Sci.*, **6,** 639.
49. Coats, A. W. and Redfern, J. P. (1964) *Nature* (London), **201,** 68; (1965) *Polym. Lett.*, **3,** 917.
50. Van Krevelen, D. W., Van Heerden, C. and Huntjens, F. J. (1951) *Fuel*, **30,** 253.
51. Horowitz, H. H. and Metzger, G. (1963) *Anal. Chem.*, **35,** 1464; (1963) *Fuel*, **42,** 418.
52. Kassman, A. J. (1980) *J. Thermal Anal.*, **18,** 199.
53. Zsakó, J. (1970) *J. Thermal Anal.*, **2,** 145; (1968) *J. Phys. Chem.*, **72,** 2406.
54. Satava, V. and Skvara, F. (1969) *J. Am. Ceram. Soc.*, **52,** 591; Satava, V. (1973) *J. Thermal Anal.*, **5,** 217.
55. Gyulai, G. and Greenhow, E. J. (1974) *J. Thermal Anal.*, **6,** 279; (1973) *Thermochim. Acta*, **6,** 239.
56. Brown, M. E. and Phillpotts, C. A. R. (1978) *J. Chem. Educ.*, **55,** 556.
57. Brown, M. E. and Galwey, A. K. (1979) *Thermochim. Acta*, **29,** 129.
58. Flynn, J. H., Brown, M. E. and Sestak, J. (1987) *Thermochim. Acta*, **110,** 101.

Chapter 14

Purity determination using DSC

14.1 Introduction

Measurements of the depression of the melting point of a sample are often used to determine its purity. Calculations are based on the assumptions that solid solutions are not formed and that the melt is an ideal solution. The practical aim of purity determinations is usually to decide whether or not the sample meets certain specifications, determined by the intended further uses of the sample.

The melting endotherm for a pure substance recorded on a DSC is illustrated in Fig. 14.1. T_0 is the melting point of the sample and the area ABC is proportional to the enthalpy of fusion, ΔH_f, of the sample. The presence of an impurity in the sample (the solvent) generally lowers the melting point of the solvent and also broadens the melting range, giving a broader DSC endotherm as illustrated in Fig. 14.2 (inset). From endotherms such as illustrated in Figs 14.1 and 14.2, melting points and enthalpies of fusion may readily be determined. With more effort an estimate of the purity of a compound can be obtained, from analysis of the detailed shape of its melting endotherm, e.g.

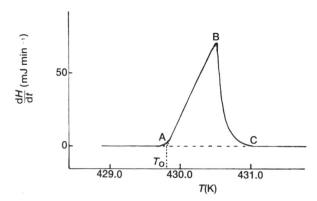

Figure 14.1 Idealized DSC record of melting of pure indium. Slope AB ($= 1/R_0$) is used to correct for thermal lag. T_0 is the melting point. Area ABC represents the enthalpy of fusion, ΔH_f°. [6]. (With the permission of the *Journal of Chemical Education*.)

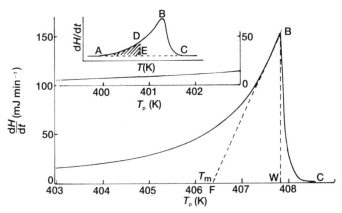

Figure 14.2 Idealized DSC record of melting of an impure sample. The slope of BF (=slope of AB in Fig. 14.1) is used to correct the programmed temperature, T_p, to the sample temperature, T_S. T_m is the melting point. The area ABC represents the enthalpy of fusion, $\Delta H^{\ominus}_{f,1}$. The fraction melted, F_n, at temperature $T_n =$ area ADE/area ABC $= a_n/A$ [6]. (With the permission of the *Journal of Chemical Education*.)

Fig. 14.2, without reference to compounds containing known amounts of impurities.

14.2 Phase equilibria

The simplest system to consider is that in which the impurity does not form a solid solution with the solvent, but forms an ideal solution in the melt, i.e. a eutectic system. If the impurity is labelled, in the customary way, as component 2 and the solvent as component 1, then for equilibrium (at constant pressure) between pure 1 in the solid and 1 in the solution (or 'melt') at activity a_1, there must be equality of the chemical potentials (μ) of 1 in the two phases:

$$\mu^{\ominus}_1(s) = \mu_1(l)$$

i.e.,
$$\mu^{\ominus}_1(s) = \mu^{\ominus}_1(l) + RT \ln a_1 \tag{14.1}$$

(where the superscript $^{\ominus}$ refers to standard conditions, i.e. unit activities). Differentiating equation (14.1) with respect to temperature, T

$$\frac{d\mu^{\ominus}_1(s)}{dT} = \frac{d\mu^{\ominus}_1(l)}{dT} + R \ln a_1 + RT \frac{d(\ln a_1)}{dT}$$

and, since $d\mu/dT = -\bar{S}$ (where the bar represents a molar quantity)

$$-\bar{S}^{\ominus}_1(s) = -\bar{S}^{\ominus}_1(l) + R \ln a_1 + RT \frac{d(\ln a_1)}{dT} \tag{14.2}$$

From equation (14.1)

$$R \ln a_1 = (\mu_1^{\ominus}(s) - \mu_1^{\ominus}(l))/T$$

so equation (14.2) on rearranging, becomes

$$\frac{d(\ln a_1)}{dT} = \frac{-[\mu_1^{\ominus}(s) - \mu_1^{\ominus}(l)]}{RT^2} - \frac{[\bar{S}_1^{\ominus}(s) - \bar{S}_1^{\ominus}(l)]}{RT}$$

$$= \frac{[\mu_1^{\ominus}(l) + T\,\bar{S}_1^{\ominus}(l)] - [\mu_1^{\ominus}(s) + T\,\bar{S}_1^{\ominus}(s)]}{RT^2}$$

or

$$\frac{d(\ln a_1)}{dT} = \frac{\bar{H}_1^{\ominus}(l) - \bar{H}_1^{\ominus}(s)}{RT^2} = \frac{\overline{\Delta H_{f,1}^{\ominus}}}{RT^2} \tag{14.3}$$

since $H = G + TS$ and $\bar{G}^{\ominus} = \mu^{\ominus}$. Integrating equation (14.3) between the limits $a_1 = 1$ at $T = T_0$ (since solid solutions are not formed) and $a_1 = a_1$ at $T = T$, assuming that $\overline{\Delta H_{f,1}^{\ominus}}$ is independent of temperature over the range,

$$\ln a_1 = \frac{\overline{\Delta H_{f,1}^{\ominus}}}{R} \left(\frac{1}{T_0} - \frac{1}{T} \right)$$

For an ideal solution, $a_1 = x_1$ (the mole fraction of 1). Hence

$$\ln x_1 = \ln(1 - x_2) = \frac{\overline{\Delta H_{f,1}^{\ominus}}}{R} \left(\frac{1}{T_0} - \frac{1}{T} \right)$$

For a dilute solution, i.e. small values of x_2,

$$\ln(1 - x_2) \simeq -x_2$$

$$x_2 = \frac{\overline{\Delta H_{f,1}^{\ominus}}}{R} \left(\frac{1}{T} - \frac{1}{T_0} \right) \tag{14.4}$$

Equation (14.4) forms the basis of melting-point depression calculations, as follows. At $T = T_m$, the melting point of the impure sample

$$x_2 = \frac{\overline{\Delta H_{f,1}^{\ominus}}}{R} \left(\frac{T_0 - T_m}{T_0 T_m} \right) = \frac{\overline{\Delta H_{f,1}^{\ominus}}}{R} \frac{\Delta T_f}{T_0 T_m} \tag{14.5}$$

If ΔT_f is small, $T_0 \simeq T_m$ and $T_0 T_m \simeq T_0^2$. Also $x_2 = n_2/(n_1 + n_2) \simeq m\,M_1/1000$, where m is the molality of solute and M_1 the molar mass of the solvent. Hence

$$\Delta T_f = \left[\frac{RT_0^2\,M_1}{\overline{\Delta H_{f,1}^{\ominus}} \cdot 1000} \right] m = K_f \cdot m$$

where K_f is termed the cryoscopic constant.

Only when the sample is completely melted, i.e. at $T \geqslant T_m$, is the mole fraction of impurity in the liquid, x_2, the same as that in the original sample, x_2^*. From the phase diagram for a simple eutectic system (Fig. 14.3) it may be seen that the value x_2^* is the minimum value which x_2 attains. At $T < T_m$ (Fig. 14.3), when the fraction of sample that has melted, F, is less than unity, the composition of the melt is closer to that of the eutectic, i.e., $x_2 > x_2^*$. When melting commences, the first liquid has the eutectic composition.

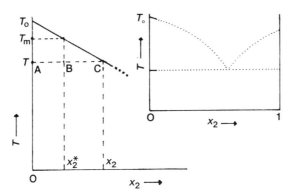

Figure 14.3 Low concentration region of a simple eutectic phase diagram (inset). By the lever rule: $BC/AB = n_{solid}/n_{liquid}$. C is the composition of the melt in equilibrium with pure solid at T. Here $x_2 > x_2^*$. $F = n_1/(n_s + n_1) = AB/(AB + BC) = AB/BC = x_2^*/x_2$. (With the permission of the *Journal of Chemical Education*.)

If F is the fraction melted at temperature T, then, assuming a linear initial segment of the liquidus curve (Fig. 14.3), and using equation (14.4)

$$F = \frac{x_2^*}{x_2} = \frac{T_0 - T_m}{T_0 - T} = \frac{x_2^* \, RT_0^2}{\overline{\Delta H_{f,1}^{\ominus}} \, (T_0 - T)} \tag{14.6}$$

Rearrangement yields

$$T = T_0 - \left[\frac{x_2^* \, RT_0^2}{\overline{\Delta H_{f,1}^{\ominus}}} \right] \frac{1}{F} \tag{14.7}$$

If F can be determined at various temperatures, T, a plot of T against $1/F$ should yield a straight line, provided that $\overline{\Delta H_{f,1}^{\ominus}}$ is independent of temperature. If the values of T_0 and $\overline{\Delta H_{f,1}^{\ominus}}$ are known, x_2^* can be determined from the measured slope of the line. The DSC curve is capable of providing values of F at temperatures T for use in such a plot.

14.3 The DSC melting curve

The DSC measures the thermal energy per unit time, dH/dt, transferred to or from the sample as the temperature of the sample holder, T, is changed at a constant rate, $dT/dt = \phi$. Thus the output from the DSC is directly proportional to the heat capacity of the system, dH/dT

$$\frac{dH}{dt} = \left(\frac{dH}{dT}\right)\left(\frac{dT}{dt}\right) \tag{14.8}$$

For an absolutely pure compound with zero melting range, dH/dt would become infinite at the melting point, T_0. For an impure compound, dH/dT is finite and is a function of T.

When the fraction melted, F, is zero, the heat capacity of the sample is that of the solid mixture, and when $F = 1$ the heat capacity of the sample is that of the ideal solution. Intermediate behaviour is obtained as follows

$$\frac{dH}{dT} = \left(\frac{dH}{dF}\right)\left(\frac{dF}{dT}\right)$$

dF/dT is obtained from equation (14.6) as

$$\frac{dF}{dT} = \frac{x_2^* \, RT_0^2}{\overline{\Delta H_{f,1}^{\ominus}} \, (T_0 - T)^2}$$

It is also assumed that, because of the restriction to consideration of ideal eutectic systems and to the formation of ideal solutions on melting, that

$$H_F = \overline{\Delta H_{f,1}^{\ominus}} F$$

and therefore

$$dH/dF = \overline{\Delta H_{f,1}^{\ominus}}$$

Combining these results

$$\frac{dH}{dT} = \frac{x_2^* \, RT_0^2}{(T_0 - T)^2} \tag{14.9}$$

Equation (14.9) then gives the variation of the heat capacity of the sample during melting as a function of T. The upper limit of the melting process is $T = T_m$ (when $F = 1$). Therefore, equation (14.7) becomes

$$T_m = T_0 - \frac{x_2^* \, RT_0^2}{\overline{\Delta H_{f,1}^{\ominus}}}$$

The lower limit of the melting process is $T \ll T_0$, when $dH/dT \simeq x_2^* \, R = C_s$ i.e. the heat capacity of the sample is approximately constant. In the idealized DSC curves given in Figs 14.1 and 14.2, it has been assumed that the heat capacity of the liquid just above the melting temperature is the same as that of the solid at lower temperatures (i.e., both equal to C_s).

Equation (14.9) can be written as

$$\frac{dH}{dT} = \frac{x_2^* \, R(T_0/T)^2}{(T_0/T - 1)^2}$$

so that, within the limits of the assumptions made above, the heat capacity during melting depends only on the mole fraction of impurity, x_2^* and the temperature relative to the melting point of the pure substance, T/T_0.

Since dH/dt is proportional to dH/dT, plots of dH/dT against T represent the initial part of an idealized DSC melting curve. Such curves for phenacetin and for benzamide, with values of x_2^* from 0.0050 to 0.3000, have been given by Marti *et al.* [1, 2].

The real DSC melting curve, because of factors such as thermal lag, which is discussed in more detail below, will look more like the curve in Fig. 14.2, inset. The total area under the curve, i.e. area ABC, is proportional to the enthalpy of fusion, $\overline{\Delta H_{f,1}^{\ominus}}$. The actual value of $\overline{\Delta H_{f,1}^{\ominus}}$ can be obtained by calibration of the instrument with a standard of known $\overline{\Delta H_f^{\ominus}}$ (Fig. 14.1). The feature sought for the present discussion, the fraction of the sample melted, F, at temperature T, is obtained directly from the fractional area under the curve, i.e. F = area ADE/area ABC. The range of F values used in practice is usually restricted to $0.1 < F < 0.4$. Even with this restricted range, the linearity of plots of T against $1/F$ (equation (14.7)) is often poor. Corrections have to be made for thermal lag and for undetected premelting as discussed below.

14.4 Corrections

14.4.1 Thermal lag

Flow of thermal energy from the holder at the programmed temperature, T_p, to the sample at a slightly lower temperature, T_s, is governed by Newton's law

$$T_p - T_s = \frac{dH}{dt} R_0 \qquad (14.10)$$

where R_0 is the thermal resistance. The value of R_0 for the instrument may be obtained [3] from the melting curve of a high purity standard (Fig. 14.1). This will melt over a very narrow temperature range, so that as the programmed temperature continues to increase with time, the sample temperature, T_s, will remain constant, i.e. $dT_s/dt = 0$. From equation (14.10)

$$\frac{dT_p}{dt} - \frac{dT_s}{dt} = R_0 \frac{d}{dt}\left(\frac{dH}{dt}\right)$$

Hence

$$\frac{dT_p}{dt} = R_0 \frac{d}{dt}\left(\frac{dH}{dt}\right) = R_0 \left(\frac{dT}{dt}\right)\left(\frac{d}{dT}\left(\frac{dH}{dt}\right)\right) = R_0 \, \phi \, \frac{d}{dT}\left(\frac{dH}{dt}\right)$$

and

$$\frac{d}{dT}\left(\frac{dH}{dt}\right) = \frac{1}{R_0}$$

R_0 may thus be determined from the slope, AB, of the DSC melting curve for the high purity standard (Fig. 14.1). The value of R_0 is then used graphically or analytically to correct the programmed temperature, T_p, to the true sample temperature, T_s. T_s then, rather than T_p, is plotted against $1/F$.

14.4.2 Undetermined premelting

Even with correction for thermal lag, the linearity of the plots of T_s against $1/F$ is often not good, and corrections have to be made to the measured areas for melting which has occurred at lower temperatures and which is difficult or impossible to measure. This is evident in the small but cumulatively significant deviation from the ideal baseline illustrated in Fig. 14.2.

If the undetermined area under the curve is ε, the measured partial areas up to temperatures T_1, T_2, ... T_n are a_1, a_2, ... a_n, respectively, and the measured total area is A, the true value of F_n is

$$F_n = (a_n + \varepsilon)/(A + \varepsilon)$$

or

$$1/F_n = (A + \varepsilon)/(a_n + \varepsilon) \simeq A/(a_n + \varepsilon),$$

since $A \gg \varepsilon$, so the effect of including ε is to reduce the value of $1/F_n$ (Fig. 14.4).

In practice ε is treated as a parameter whose value is adjusted so that a plot of T_s against the corrected $1/F$ is linear. The restraints are that the final value of $(A + \varepsilon)$ should correspond directly to the correct value of $\overline{\Delta H_{f,1}^{\ominus}}$ (if known) and that the value of T_0, determined from the intercept on the T_s axis, should be correct. Once these conditions have been met, the value of $x_2^* \ (= \text{slope} \times \overline{\Delta H_{f,1}^{\ominus}}/RT_0^2)$ and hence the purity of the sample can be

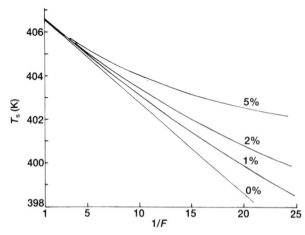

Figure 14.4 Correction for undetermined premelting. The correction $= 100 \times \varepsilon/A$ and $F = (a + \varepsilon)/(A + \varepsilon)$. (With the permission of the *Journal of Chemical Education*.)

determined. Since melting actually begins at the eutectic temperature which may be far below the range of temperatures being examined, the correction, ε, may sometimes be quite large and values of as much as 30% of the total area are not uncommon [9]. Obviously the approximation, $A + \varepsilon \simeq A$, cannot then be used. Sondack [5] has suggested an alternative procedure.

The whole procedure for purity determination is summarized in the flowchart (Fig. 14.5). Computer programs for these calculations have been described [1, 2, 4]. More detail on the procedure is given in ref. 5.

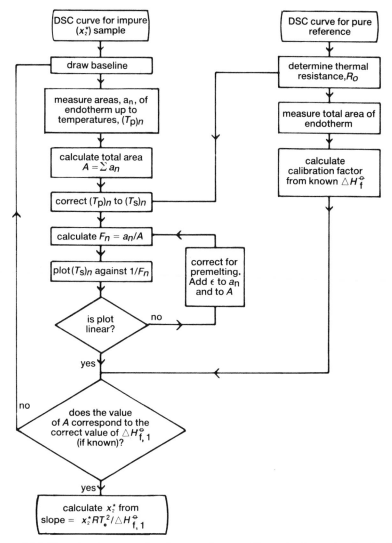

Figure 14.5 Flowchart representing the procedure used in purity determination by DSC [6]. (With the permission of the *Journal of Chemical Education*.)

14.5 Step methods

Staub and Perron [7] have shown that a stepped heating technique can extend the working region for purity determination. The sample is heated through the melting region in steps of a few tenths of a degree, thus allowing a closer approach to true thermodynamic equilibrium. The results of such a procedure [8] on an impure phenacetin sample, using a modified DSC-1B are shown in Fig. 14.6. There are 6 steps/5 K and the areas of each peak, obtained by integration, are given. The first 11 peaks and the last 3 arise from the difference in the heat capacity of the sample and reference. These peaks have approximately constant area and the melting peaks can be corrected for this difference. Because thermal equilibrium is established after each peak, no correction is required for the thermal resistance, R_0, of the system. The corrected areas of all the melting steps are then summed and converted to fractional areas, F, and $1/F$ is plotted against T, as before, except that with this procedure [8] all the data are used including the latter part of the melting process. The linearization process for the undetermined premelting still has to be carried out.

Gray and Fyans [9] have suggested an alternative procedure to that given above, in which the mole fraction of impurity (x_2^*) is calculated from the areas of two consecutive stepped peaks (α_n and α_{n-1}), the magnitude of the stepping interval (ΔT), and the molar mass (M) and the melting point (T_0) of the pure major component. This method depends upon the applicability of the van't Hoff equation and the relationship derived [9] is

$$x_2^* = \frac{2M}{RT_0^2} \alpha_n \alpha_{n-1} \Delta T \frac{(\alpha_n + \alpha_{n-1})}{(\alpha_n - \alpha_{n-1})^2}$$

The melting point (T_0) of the pure solvent may be determined [9] from the areas α_n and α_{n-1}, the step interval ΔT and the final temperature of the step T_n

$$T_0 = T_n + 2\Delta T \frac{(\alpha_{n-1})}{(\alpha_n - \alpha_{n-1})}$$

It is recommended [9] that as large step increments as possible should be used.

14.6 Conclusions

Raskin [10], in a critical review of methods of purity determination using DSC, concludes that the accuracy of the method is generally overestimated. A realistic estimate of the accuracy in the range $0.005 < x_2^* < 0.02$ is given [10] as 30–50% when the sample mass is less than 3 mg and the scanning rate is less than 2 K min^{-1}.

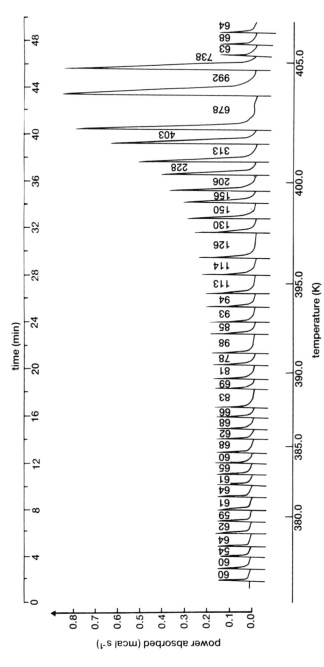

Figure 14.6 Stepwise heating of an impure (96.04 mole %) phenacetin sample [8]. Peak areas are expressed in thousands. (With the permission of the American Chemical Society.)

Garn *et al.* [11–13] have discussed the problems that arise when there is appreciable solid solubility. They used NMR to detect the solidus and compared DSC and NMR results. They showed that lack of thermal equilibrium is not a principal source of error in the method and also found that the measured impurity content is sometimes dependent upon the nature of the impurity as well as its concentration.

14.7 References

1. Marti, E. E. (1973) *Thermochim Acta*, **5**, 173.
2. Marti, E. E., Heiber, O., Huber, W. and Tonn, G. (1971) *Thermal Analysis*, Proc. 6th ICTA, Birkhauser Verlag, Basel, Vol. 3, p. 83.
3. *Thermal Analysis Newsletters*, Nos. 5 and 6, Perkin-Elmer Corporation, Norwalk, Connecticut (undated).
4. Moros, S. A. and Stewart, D. (1976) *Thermochim. Acta*, **14**, 13.
5. Sondack, D. L. (1972) *Anal. Chem.*, **44**, 888.
6. Brown, M. E. (1979) *J. Chem. Educ.*, **56**, 310.
7. Staub, H. and Perron, W. (1974) *Anal. Chem.*, **46**, 128.
8. Zynger, J. (1975) *Anal. Chem.*, **47**, 1380.
9. Gray, A. P. and Fyans, R. L. (1973) *Thermal Analysis Application Study No. 10*, Perkin-Elmer, Norwalk.
10. Raskin, A. A. (1985) *J. Thermal Anal.*, **30**, 901.
11. Garn, P. D., Kawalec, B., Houser, J. J. and Habash, T. F. (1982) *Proc. 7th ICTA*, Vol. 2, Wiley, Chichester, p. 899.
12. Kawalec, B., Houser, J. J. and Garn, P. D. (1982) *J. Thermal Anal.*, **25**, 259.
13. Habash, T. F., Houser, J. J. and Garn, P. D. (1982) *J. Thermal Anal.*, **25**, 271.

Chapter 15

The literature and nomenclature of thermal analysis

A selection of references is given below to serve as an introduction to the vast literature of thermal analysis. A more comprehensive list is given in *For Better Thermal Analysis*, published by ICTA, and soon to come out in a new edition by Dr J. O. Hill.

15.1 Books

Blazek, A. (1972) *Thermal Analysis*, Van Nostrand Reinhold, London.

Daniels, T. (1973) *Thermal Analysis*, Kogan Page, London.

Duval, C. (1962) *Inorganic Thermogravimetric Analysis*, 2nd Rev. Edn., Elsevier, Amsterdam.

Earnest, C. M. (1984) *Thermal Analysis of Clays, Minerals and Coals*, Perkin-Elmer, Norwalk.

Garn, P. D. (1965) *Thermoanalytical Methods of Investigation*, Academic Press, New York.

Heide, K. (1982) *Dynamic Thermal Analysis Methods* (in German), 2nd Edn. Deutscher Verlag für Grundstoffindustrie, Leipzig.

Keattch, C. J. and Dollimore, D. (1975) *An Introduction to Thermogravimetry*, 2nd Edn., Heyden, London.

Lodding, W. (Ed.) (1976) *Gas Effluent Analysis*, Arnold, London.

Lu, C. and Czanderna, A. W. (Eds.) (1984) *Applications of Piezoelectric Quartz Crystal Microbalances*, Elsevier, Amsterdam.

Mackenzie, R. C. (Ed.) (1969) *Differential Thermal Analysis*, Vol. 1 and 2, Academic Press, London.

McNaughton, J. L. and Mortimer, C. T. (1975) *Differential Scanning Calorimetry*, Perkin-Elmer Order No.: L-604 (Reprinted from IRS: Phys. Chem. Ser. 2, Vol. 10, Butterworths).

Marti, E., Oswald, H. R. and Wiedemann, H. G. (Eds.) (1979) *Angewandte chemische Thermodynamik and Thermoanalytik*, Birkhauser Verlag, Basel.

Menis, O., Rook, H. L. and Garn, P. D. (Eds.) (1980) *The State-of-the-Art of Thermal Analysis*, NBS Special publication 580.

Mikhail, R. Sh. and Robens, E. (Eds.) (1982) *Microstructure and Thermal Analysis of Solid Surfaces*, Wiley, Chichester.

Pope, M. I. and Judd, M. D. (1977) *Differential Thermal Analysis*, Heyden, London.

Todor, D. N. (1976) *Thermal Analysis of Minerals*, Abacus Press, Tunbridge Wells.

Turi, E. A. (Ed.) (1981) *Thermal Analysis in Polymer Characterization*, Heyden, Philadelphia.

Turi, E. A. (Ed.) (1981) *Thermal Characterization of Polymeric Materials*, Academic Press, New York.

Wendlandt, W. W. (1986) *Thermal Analysis*, 3rd Edn., Wiley, New York.

Wendlandt, W. W. and Collins, L. W. (Eds.) (1976) *Thermal Analysis* (Benchmark Papers in Analytical Chemistry), Dowden, Hutchinson & Ross, Stroudsbourg, USA.

Wendlandt, W. W. and Smith, J. P. (1967) *Thermal Properties of Transition Metal Ammine Complexes*, Elsevier, Amsterdam.

Wilson, C. L. and Wilson, D. W. (Eds.), *Comprehensive Analytical Chemistry*, Elsevier, Amsterdam.

(1981) Paulik, J. and Paulik, F., *Simultaneous Thermoanalytical Examinations by Means of the Derivatograph*, Vol. XIIA.

(1982) Jesperson, N. D. (Ed.), *Biochemical and Clinical Applications of Thermometric and Thermal Analysis*, Vol. XIIB.

(1983) Balek, V. and Tolgyessy, J., *Emanation Thermal Analysis and other Radiometric Emanation Methods*, Vol. XIIC.

(1984) Sestak, J., *Thermophysical Properties of Solids*, Vol. XIID.

Wolsky, S. P. and Czanderna, A. W. (Eds.) (1980) *Microweighing in Vacuum and Controlled Environments*, Elsevier, Amsterdam.

15.2 Reviews

An immediate introduction to the literature is provided by the biennial reviews in *Analytical Chemistry*, (1986) **58**, 1R; (1984) **56**, 250R; (1982) **54**, 97R; (1980) **52**, 106R; (1978) **50**, 143R; (1976) **48**, 341R; (1974) **46**, 451R; (1972) **44**, 513R; (1970) **42**, 268R; (1968) **40**, 380R; (1966) **38**, 443R; (1964) **36**, 347R and earlier reviews on DTA.

Many other more specialized reviews have appeared in the Journals mentioned in section 15.4.

15.3 Conference proceedings

The proceedings of the International Conferences on Thermal Analysis (ICTA) and the European Symposium on Thermal Analysis and Calorimetry (ESTAC) are valuable surveys of the 'state-of-the-art' (Tables 15.1 and 15.2).

The annual conferences of the North American Thermal Analysis Society (NATAS) are published in short form and are very useful.

A series on *Analytical Calorimetry* based on the ACS Symposia, edited by R. S. Porter and J. F. Johnson, is published by Plenum Press: Vol. 1, 1968; Vol. 2, 1970; Vol. 3, 1974, Vol. 4, 1978 and Vol. 5, 1984.

The series *Progress in Vacuum Microbalance Techniques*, published by Heyden, and the earlier volumes *Vacuum Microbalance Techniques*, published by Plenum Press, are sources of information on TG.

Table 15.1 International Conferences on Thermal Analysis (ICTA)

ICTA	*Date*	*Place*	*Editor*	*Publisher*
8th	1985	Czechoslovakia	A. Blazek	*Thermochimica Acta* Vols 92, 93, 110
7th	1982	Canada	B. Miller	Wiley
6th	1980	Germany	H. G. Wiedemann	Birkhauser Verlag
5th	1977	Japan	H. Chihara	Heyden
4th	1974	Hungary	I. Buzas	Heyden
3rd	1971	Switzerland	H. G. Wiedemann	Birkhauser Verlag
2nd	1968	United States	R. F. Schwenker Jr. and P. D. Garn	Academic Press
1st	1965	United Kingdom	J. P. Redfern	Macmillan

Table 15.2 European Symposia on Thermal Analysis and Calorimetry (ESTAC)

ESTAC	*Date*	*Place*	*Editor*	*Publisher*
3rd	1984	Switzerland	B. F. Rordorf	*Thermochimica Acta* Vol. 85 (1985) 1–527
2nd	1981	United Kingdom	D. Dollimore	Heyden
1st	1976	United Kingdom	D. Dollimore	Heyden

15.4 Journals

The *Journal of Thermal Analysis* (Wiley) and *Thermochimica Acta* (Elsevier) are the main specialist journals, although results of thermal analyses appear in many other journals, especially *Analytical Chemistry*, *Talanta*, *Analytica*

Chimica Acta, International Laboratory, Laboratory Practice, Analyst, and the many polymer journals. The *Journal of Thermal Analysis* has a useful bibliography section and Wiley also publishes *Thermal Analysis Abstracts* (Fig. 15.1).

Computer-based searching services, e.g., *CA Selects: Thermal Analysis,* published by the American Chemical Society, are a convenient way of keeping abreast of the literature (Fig. 15.2).

15.5 Manufacturer's literature

The major manufacturers (listed in chapter 16) provide extensive information on thermal analysis. Special mention must be made of the *Thermal Analysis Newsletters, Thermal Analysis Application Studies, Instrument News* and bibliographies produced by Perkin-Elmer; the *Applications Briefs* and *Du Pont Thermogram;* The *Cahnsultant; The Technical Information Sheets* from Stanton Redcroft, and the *Thermal Techniques Series* from Mettler. The manufacturers often provide reprints of articles, published in the literature, but based on their equipment.

15.6 The nomenclature of thermal analysis

The International Confederation for Thermal Analysis (ICTA) has published several recommendations for the standardizing and reporting of results of thermal analysis. Many of these are contained in a booklet *For Better Thermal Analysis* published by ICTA. Other references are: Mackenzie, R. C. (1969) *Talanta,* **16,** 1227; (1974) *Pure Appl. Chem.,* **37,** 439; (1972) *Talanta,* **19,** 1079; (1975) *J. Therm. Anal.,* **8;** (1979) *Thermochim. Acta.,* **28,** 1 and (1981) **46,** 333. See also *The Metrication of Thermal Analysis or Conversion to SI Units,* Blaine, R. L. (1978) *Thermochim. Acta,* **26,** 217–228.

Thermal Analysis Abstracts

An official organ of the International Confederation for Thermal Analysis

Sample Abstract

Abstract Number	13-1250
Title	TEMPERATURE DEPENDENCE OF THE HEAT-CAPACITY CHANGE FOR THE DISSOCIATION OF ACETIC ACID AND OF PROPIONIC ACID IN WATER
Author(s)	Olofsson, G.
Address	Thermochemistry Laboratory, Chemical Centre, P.O. Box 740, S-220 07 Lund, Sweden.
Reference (Language)	J. Chem. Thermodyn., 16 (1), 39-44. 1984 (Eng).
Abstract	The molar enthalpy of proton dissociation, determined at various temperatures between 274 and 373 K, varied linearly with temperature for both acids. The two acids gave virtually the same heat capacity, the average value of $\triangle_a C_{p.m}^{\circ}$ being -142 J/K.mol.

Keyword Numbers	010	109, 131, 136	220, 227	319, 595
	Methodology	Technique/ Application	Materials/ Subjects	Elements/ Compounds

Abstractor	CS

KEY	010 — the method used was calorimetry
	109 — determination of heat of reaction, etc
	131 — determination of specific heat or heat capacity
	136 — determination of thermodynamic constants
	220 — the materials investigated were liquid
	227 — the materials investigated were organic
	319 — the materials investigated involved organic acids or salts
	595 — the materials investigated involved water

Figure 15.1 Sample entry from *Thermal Analysis Abstracts*. (With the permission of John Wiley & Sons, Chichester.)

```
ITEM NO 013127        CA VOL 105 NO 02      * PROFILE 0637

AUTHORS: WENDLANDT, W. W.(*)
         (*) DEP. CHEM., UNIV. HOUSTON, HOUSTON, TX, 77004,
    USA

TITLE: THE DEVELOPMENT OF THERMAL ANALYSIS INSTRUMENTATION
    1955-1985

REFERENCE: THERMOCHIM. ACTA (THACA) VOL 100 NO 1 PP 1-22
    (1986)    ISSN 00406031
INDEX TERMS:
    SECTIONS: CA069000 THERMODYNAMICS, THERMOCHEMISTRY
       CA020; CA073; CA079; CA080
    DESCRIPTORS:
       THERMAL ANALYSIS, APP.... (DEVELOPMENT OF)
    IDENTIFIERS:
       REVIEW THERMAL ANALYSIS INSTRUMENT, DTA INSTRUMENT REVIEW,
       THERMOELECTROMETRY INSTRUMENT REVIEW, LIGHT EMISSION
       INSTRUMENT THERMAL ANALYSIS REVIEW, THERMOBALANCE
       INSTRUMENT THERMAL ANALYSIS REVIEW

SEARCH TERMS PRESENT: THERMAL ANAL ; CA079

                              NO ABSTRACT GIVEN
```

Figure 15.2 Sample output from a computer-based literature searching service covering *Chemical Abstracts*. (National Institute for Informatics, CSIR, Pretoria.)

Chapter 16

Thermal analysis equipment

16.1 Choosing thermal analysis equipment

A bewildering array of equipment is available to anyone starting out in thermal analysis and selecting a suitable system is not an easy decision. It is advisable first to collect the pamphlets and specifications of most systems from the suppliers (section 16.2), and then attempt to make a preliminary selection by considering the following factors.

(1) What types of sample are going to be examined both immediately and as far as can be predicted, in the future?
(2) What sort of information on the sample is required, e.g. thermal stability, glass-transitions, percentage crystallinity, mechanical properties, details of gases evolved etc.?
(3) Over what temperature ranges are the changes that are of interest likely to occur?

Answers to the above questions can be used to eliminate definitely unsuitable systems. Unless there are very special requirements such as extreme ranges of temperature, or use of very corrosive atmospheres or strongly exothermic or even explosive samples, there will usually still be a wide choice of instruments. Prices related to budget available will obviously be a further restriction, and modular systems, which can be added to, are attractive. Most systems incorporate their own computers and it is worth looking carefully at this (chapter 12) to see whether a system can be interfaced to existing laboratory computers. This should only be contemplated if you have a source of software. Some systems cannot operate without the manufacturer's computer.

A most important question to be answered before choosing a system, is the training and service available from the suppliers (section 16.2). This involves seeking out other users and checking on their experiences. At the same time it is worth checking on the prices of spares and accessories (which can be ridiculously high). Even the best equipment (which need not be the most expensive) will be difficult to operate and maintain if the local agents cannot provide informed and rapid service.

16.2 Major suppliers of thermal analysis equipment

Manufacturer	Address
Cahn-Ventron Corporation	16207 South Carmenita Road, Cerritos, CA 90701, USA.
DuPont Company Instrument Systems	Concord Plaza, Quillen Building, Wilmington, DE 19898, USA.
Harrop	Harrop Laboratories, 3470 East Fifth Avenue, Columbus, Ohio 43219, USA.
Linseis Messgerate GmbH	Vielitzerstrasse 43, 8672 Selb, FRG.
Maple Instruments	Pissummerweg 1, 6114 AH Susteren, The Netherlands.
Mettler Instrumente AG	CH-8606 Greifensee, Switzerland.
	P.O. Box 71, Hightstown, NJ 08520, USA.
Netzch-Geratebau GmbH	P.O. Box 1460, D-8672 Selb, FRG.
Perkin-Elmer Corporation Analytical Instruments	Main Avenue (MS-12), Norwalk, Connecticut 06856, USA.
Polymer Laboratories	Essex Road, Church Stretton, SY6 6AX, England.
Rigaku-Denki Co.	9-8, 2-Chome, Sotokanda, Chiyoda-Ku, Tokyo 101, Japan.
Setaram	7, rue de l'Oratoire, 69300 Caluire, France.
Shimadzu Corporation	Shinjuku Mitsui Bldg, 1-1, Nishi-Shinjuku 2-Chome, Shinjuku-ku, Tokyo 160, Japan.
Stanton Redcroft	Copper Mill Lane, London SW17 0BN, England.

Chapter 17

Conclusion

In the preceding sections, only a few selected examples have been given of the vast number of applications of thermal analysis. To get a better idea of the range of applications over all branches of inorganic, organic, physical and industrial chemistry, metallurgy, polymer science, glass and ceramic science, food science, biology, geology, pharmacy, medicine, agriculture and engineering, it is worth scanning the contents pages of some of the proceedings of the International Conferences on Thermal Analysis (ICTA), as well as the biennial reviews by W. W. Wendlandt in *Analytical Chemistry* (detailed references are given in chapter 15). Although applications have been grouped under the main techniques used, many of the studies reported have been, or could have been, extended by the use of simultaneous and/or complementary measurements (chapter 9).

As suggested in the opening paragraphs of this book, there are very few materials which will not show interesting changes on heating. The materials studied have ranged from kidney stones to synthetic diamonds, and from ancient papyri to the latest polymers and ceramics for space research. The information obtained has been used to solve old problems and to develop new processes for the future.

Appendix A

Introductory experiments in thermal analysis

A selection of introductory experiments is given below. What can be done will obviously be determined by the apparatus available, and constant reference should be made to the manuals supplied with the instruments. Modern computerized instruments will usually operate interactively, providing some of the directions given in the procedures below.

A.1 Differential scanning calorimetry (DSC)

A.1.1 Calibration

Calibrate the DSC with respect to temperature and heat flow. Check on the reproducibility. Compare the results obtained on different instruments, if possible.

A.1.2 Determination of peak areas

Use a planimeter to determine areas recorded on chart paper. Use computer programs to carry out numerical integration.

A.1.3 Dehydration

Determine the temperatures and enthalpy changes for the dehydration stages of $CuSO_4 . 5H_2O$.
 Carry out similar measurements on some other hydrates, e.g. $BaCl_2 . 2H_2O$, $Ni(HCOO)_2 . 2H_2O$ or $Ni(COO)_2 . 2H_2O$ and see how the enthalpy of dehydration per mole of H_2O varies from salt to salt.

A.1.4 Decompositions

Study the decompositions of some metal carboxylates, e.g., $Ni(HCOO)_2 . 2H_2O$ or $Ni(COO)_2 . 2H_2O$, after first having dehydrated them (see section A.1.3). Try to carry out an isothermal DSC run at a suitable temperature. Carry out kinetic measurements.

A.1.5 Phase transitions

Determine the temperatures and enthalpy changes of phase transitions in salts such as NH_4NO_3, KNO_3, $KClO_4$ or Ag_2SO_4. Study the reversibility of these changes on cooling and comment on the use of these transition temperatures as temperature standards for instrument calibration. Use a hot-stage microscope (HSM) for visual detection of the phase changes.

A.1.6 Specific heat capacity

Determine the specific heat capacity of an inert substance, such as Al_2O_3, relative to that of aluminium metal $(0.900\ J\ K^{-1}g^{-1}$ at 25°C). Comment on the variation of the specific heat capacity with temperature.

A.1.7 Purity determination

Compare the melting endotherm of pure indium with that for benzoic acid and estimate the purity of the benzoic acid.

A.1.8 Polymer stability

Do runs on some polymer samples in nitrogen and in oxygen. Examine the results to determine the glass-transition temperatures and the temperatures at which melting, degradation and oxidation occur.

A.1.9 Polymer crystallinity

Polyethylene (PE) is a semicrystalline thermoplastic. From a DSC curve, determine the temperature range over which melting occurs as well as the enthalpy of melting. The percentage crystallinity is calculated by comparing the measured value with that for 100% crystalline material $(290\ J\ g^{-1})$.

A.1.10 Curing of an epoxy resin

Do a DSC run on epoxy resin. Determine the enthalpy change and the kinetics of the exothermic curing process. Rescan the product and get the glass-transition temperature of the polymer.

A.2 Thermogravimetry (TG)

A.2.1 Temperature calibration

Use magnetic standards of known Curie point, to calibrate the furnace temperature.

A.2.2 Dehydration

Determine the temperatures and mass losses accompanying the dehydration stages of $CuSO_4 . 5H_2O$. Compare your results with those from the DSC experiment A.1.3 and draw up a full description of the dehydration process.

A.2.3 Decompositions

The studies carried out in the DSC experiment A.1.4 can be complemented by determining the mass losses. Kinetics may be determined from a non-isothermal run or from a series of isothermal runs.

A.2.4 Percentage filler in a polymer

Decompose a sample of an epoxy putty in nitrogen. The residue is the filler.

A.2.5 Analysis of coal or of a rubber

(a) Heat a sample in nitrogen until no further mass loss occurs. This gives the proportion of volatile material.
(b) Change the purge gas to oxygen while holding the sample at the high end of the temperature range. The mass loss corresponds to the proportion of carbon residue.
(c) The mass of the residue in oxygen corresponds to the inorganic ash.

Appendix B

Thermal analysis software

Listings of several programs, written in Applesoft BASIC by the author, for use on APPLE II PLUS microcomputers with at least 48K, are given below. Although every care has been taken, no responsibility is assumed for the correct operation of these programs. Copies of the programs may be obtained by sending a blank diskette to the author. The programs should in most cases be fairly readily adaptable to other microcomputers.

The sequence of operations in the capture and processing of data from thermal analysis experiments is usually:

B.1 Data capture and temporary storage;
B.2 Permanent storage;
B.3 Display of data for preliminary examination;
B.4 Modification of the data; (a) baseline correction; (b) smoothing; (c) scaling;
B.5 Processing of the data; (a) numerical differentiation; (b) peak integration;
B.6 Other calculations, e.g., (a) kinetic analysis; (b) purity determination, etc.

B.1 and 2 Data capture and storage

Program DSC is given here as an example. This program has been designed for operation with a Perkin-Elmer DSC2 but is readily adaptable to other instruments and to other TA techniques. The program given here is based on the use of a Datel Systems Inc., ADC-ET12 12-bit analogue-to-digital converter (ADC) (see program lines 315 to 330) and a Mountain Hardware clockboard (see lines 175 to 314). Full details of the hardware are available from the author.

> Input: ADC channel number, filename for data storage, starting and stopping temperatures, heating rate and number of points to be collected.
>
> Output: Temperature and DSC response stored in a sequential text file. Data is plotted on the screen and may be dumped to a printer (Epson).

```
1   REM     CHECK THAT PROGRAM IS LOADED ABOVE SCREEN#2

2   IF  PEEK (104) = 96 THEN 5
3   POKE 103,1: POKE 104,96: POKE 4096 * 6,0
4   PRINT  CHR$ (4)"RUN DSC"
5   HOME
6   PRINT "*****************************************"
7   PRINT "* PROGRAM DSC          29/05/87 MEB   *"
8   PRINT "*****************************************"
10  PRINT : PRINT
15  AD = 5:SLOT = 2: REM  AD =A/D SLOT NO., AND SLOT=C
    LOCK SLOT NO.
20  PRINT "PROGRAM TO CAPTURE DATA FROM THE PERKIN-EL
    MER DSC2 ": PRINT
25  PRINT "THIS VERSION USES THE MEDIUM SPEED 12 BIT
    A/D BOARD SUPPLIED BY RHODES ELECTRONICS AND PUT
    IN SLOT#";AD;" SET IN LINE 15, AND THE MOUNTAIN C
    LOCK IN SLOT#";SLOT;" ALSO SET IN LINE 15 ": PRINT

26  PRINT
27  REM --------------------------INPUT---------------
29  INVERSE
30  INPUT "ENTER A/D CHANNEL NO. 0 TO 3   ";C
31  IF C < 0 OR C > 3 THEN 30
36  PRINT : INVERSE : INPUT "DO YOU WANT TO STORE THE
     DATA IN A DISK FILE ? Y OR N ? ";K$
37  IF K$ <  > "Y" THEN 39
38  PRINT : INPUT "ENTER FILENAME ";G$: PRINT
39  HOME
40  PRINT : INPUT "ENTER STARTING TEMP/KELVIN ";KO
41  IF KO < 300 THEN  PRINT "TOO LOW ! ":: GOTO 40
42  PRINT : INPUT "ENTER STOPPING TEMP/KELVIN ";KF
43  IF KF > 750 THEN  PRINT "TOO HIGH ! ": GOTO 42
44  PRINT : INPUT "ENTER HEATING RATE/K PER MIN, ENTE
    R 0 FOR ISOTHERMAL ";HR
45  IF HR < 0 OR HR > 320 THEN  PRINT "OUTSIDE SETTIN
    GS AVAILABLE ": GOTO 44
46  IF HR <  > 0 THEN 50
47  PRINT : INPUT "ENTER TIME INTERVAL BETWEEN ISOTHE
    RMAL READINGS, IN SECS ";H
50  PRINT : INPUT "ENTER NO.OF POINTS TO BE COLLECTED
     N.B. MAXIMUM IS 500 ";W: PRINT
60  IF W > 500 THEN 50
61  IF K$ <  > "Y" THEN 64
62  INPUT "IS THE RIGHT DISK IN DRIVE 1 FOR DATA STOR
    AGE ? Y OR N ? ";Y$
63  IF Y$ <  > "Y" THEN 62
64  PRINT : IF HR = 0 THEN 68
65 H = 60 * (KF - KO) / (HR * W)
66  REM  H = TIME INTERVAL = 60 X T RANGE / ( HEATING
    RATE X NO. OF POINTS )
```

```
68   IF H < 0.5 THEN  PRINT "TIME INTERVAL BETWEEN COL
     LECTION OF POINTS IS TOO SHORT. REENTER ": GOTO 4
     7
70   DIM R(W + 5),V(W + 5)
71   DIM T(W + 5)
90   D$ =  CHR$ (4): REM  "CONTROL D"
100   PRINT D$;"NOMONI,O,C": REM -KEEP DISK COMMANDS F
     ROM PRINTING
110   HOME : REM -CLEAR SCREEN
120  I = O: REM  - FLAG FOR START AND INITIALISE COUNT
     ER
129   NORMAL
130   PRINT "POINT    TIME/S","TEMP/K","DSC OUTP"
150   PRINT "-------------------------------------------":
     IF X = 1 THEN 510
151   REM  - X IS USED AS THE FLAG FOR PRINTOUT. IT IS
     SET IN LINE 500
159   PRINT : FLASH
160   INPUT "READY TO START ? Y OR N ? ";B$
161   NORMAL
170   IF B$ <  > "Y" THEN 160
175   REM ------------------CLOCK----------------
180   PRINT D$;"IN#";SLOT: REM -SETS INPUT TO CLOCK
190   PRINT D$;"PR#";SLOT: REM  -SETS OUTPUT TO THE CL
     OCK
200   INPUT " ";T$: REM -OBTAIN THE TIME
202  H$ =  MID$ (T$,7,2)
203   REM   H$ CONTAINS THE HOUR PART OF T$
210  M$ =  MID$ (T$,10,2)
211   REM   M$ CONTAINS THE "MINUTES" PART OF T$
220  S$ =  MID$ (T$,13,2)
221   REM   S$ CONTAINS THE INTEGER "SECONDS" PART OF T
     $
230  F$ =  RIGHT$ (T$,3)
231   REM   F$ CONTAINS THE FRACTIONAL PART OF "SECONDS
     " FROM T$
240  S$ = S$ + "." + F$
241   REM   INTEGER AND FRACTIONAL PARTS OF "SECONDS" A
     RE COMBINED WITH A DECIMAL POINT
250  T = 3600 *  VAL (H$) + 60 *  VAL (M$) +  VAL (S$)

251   REM    CONVERTS HOURS AND MINS -->SECONDS AND AD
     DS SECONDS
260   IF I <  > O THEN 280
261   REM  - CHECKS FLAG FOR START OR MIDDLE OF RUN
270  TO = T:Q$ = T$
271   REM  - STORES STARTING TIME
272  T(I) = O: REM  INITIALISE RUN TIME
280  R = T - TO
281   REM  - MEASURES ELAPSED TIME
290   IF I = O THEN 310
```

```
300   IF (R - I * H) < O THEN 180
301   REM    - CHECK WHETHER TIME INTERVAL HAS ELAPSED
      N.B.H=TIME INTERVAL IN SECONDS CALCULATED IN LI
      NE 65
305 T(I) = T(I) + R
310 R(I) = (R * HR / 60) + KO
311 R(I) =  INT (1000 * R(I)) / 1000: REM    3 DEC PLA
    CES
312   PRINT D$;"IN#0"
314   PRINT D$;"PR#0"
315   REM ------------------A / D---------------
320   POKE  - 16254 + AD * 16,240: POKE  - 16253 + AD *
      16,0
325   POKE  - 16256 + AD * 16,C * 32: POKE  - 16244 +
      AD * 16,12
326   FOR N = 1 TO 20: NEXT N
330 V(I) = ( PEEK ( - 16256 + AD * 16) * 256 +  PEEK
      ( - 16255 + AD * 16)) - (C * 8192): REM    READ CHA
      NNEL C
355 T(I) =  INT (T(I) * 1000) / 1000:R(I) =  INT (R(I
    ) * 1000) / 1000:V(I) =  INT (V(I) * 1000) / 1000

360   PRINT I;"    ";T(I),R(I),V(I): PRINT
370 I = I + 1: REM   - COUNTER
380   IF I = W + 1 THEN 400: REM  N.B. W=NO. OF POINTS
      TO BE COLLECTED, SET IN LINE 50
390   GOTO 180
400   PRINT : PRINT "STARTING TIME = ";TO,Q$
410   PRINT : PRINT "STOPPED AT T = ";T,T$
428   REM ------------------OUTPUT---------------
429   INVERSE
430   PRINT : INPUT "DO YOU WANT A PRINTOUT ? Y OR N ?
      ";A$
440   IF A$ = "N" THEN 510
450   PRINT : INPUT "IS PRINTER ON ? Y OR N ? ";A$
451   NORMAL
460   IF A$ <  > "Y" THEN 450
470   PR# 1
480   PRINT : PRINT "OUTPUT FROM DSC "
482   IF K$ <  > "Y" THEN 500
484   PRINT "STORED IN FILE ";G$
486   PRINT "STARTING TEMP/K = ";KO: PRINT "STOPPING T
      EMP/K = ";KF: PRINT "HEATING RATE / K PER MIN ";H
      R: PRINT
490   PRINT "=================================="
500   PRINT :X = 1: GOTO 130
501   REM  PUTTING X=1 SETS FLAG FOR PRINTOUT, SEE LIN
      ES 150 AND 420
510 MAX = 0:B = V(0)
520   FOR I = 0 TO W
530   IF A$ = "N" THEN 550
```

```
540   PRINT : PRINT I;"    ";T(I),R(I),V(I)
550   IF V(I) < MAX THEN 570
560 MAX = V(I)
570   IF V(I) > B THEN 590
580 B = V(I)
590   NEXT I
599   NORMAL
600   PRINT : PRINT "MAX = ";MAX,"MIN = ";B
610   IF A$ = "N" THEN 652
620   PRINT : PRINT "DATA FROM CHANNEL ";C
630   PRINT : PRINT "READ AT TIME INTERVALS OF ";il;" S
      ECS": PRINT
650   PR# 0
651   PRINT : FLASH : PRINT "WHEN PRINTER HAS FINISHED
      , PRESS ANY KEY TO CONTINUE ": NORMAL : GET V$
652   IF K$ < > "Y" THEN 659
654   GOTO 1590
658   REM ------------------PLOT------------------
659   INVERSE
660   PRINT : INPUT "DO YOU WANT A PLOT OF YOUR READIN
      GS ? Y OR N ? ";A$
661   NORMAL
670   IF A$ < > "Y" THEN 1990
675   IF HR = 0 THEN D1 = T(W): GOTO 685
680 D1 = KF - KO
685 RANGE = MAX - B
690 Y9 = MAX + 0.05 * RANGE:YO = B - 0.05 * RANGE
700 D2 = Y9 - YO
720   IF D2 < = 0 THEN 940
730   REM -OUTLINE FRAME
740   HGR
745   HCOLOR= 3
750   HPLOT 9,0 TO 279,0 TO 279,150 TO 9,150 TO 9,0
760   REM -TIC INTERVALS
770 G4 =   INT (270 / 10)
780 G5 =   INT (150 / 10)
790   FOR K = 1 TO 9
800   HPLOT 9 + K * G4,150 TO 9 + K * G4,147
810   HPLOT 9,147 - K * G5 TO 12,147 - K * G5
820   NEXT K
830   REM -PLOTTING OF THE POINTS
840   FOR I = 0 TO W
845   IF HR = 0 THEN X =   INT (270 * T(I) / D1) + 9: GOTO
      860
850 X =   INT (270 * (R(I) - KO) / D1) + 9
860 Y = 150 -   INT (150 * (V(I) - YO) / D2)
865   IF X < 0 OR X > 279 OR Y < 0 OR Y > 150 THEN 900
870   IF I = 0 OR V = 0 OR U = 0 THEN   HPLOT X,Y: GOTO
      890
880   HPLOT V,U TO X,Y
890 V = X:U = Y
```

```
900    NEXT I
910    PRINT
920    PRINT "X TIC = ";D1 / 10;" Y TIC = ";D2 / 10
930    GOTO 1000
940    PRINT "Y MAX - Y MIN IS ZERO   PLOT ABANDONED"
950    GOTO 2000
1000   GOSUB 7000
1010   IF Z$ <  > "Y" THEN 2000
1141   PR# 1
1142   PRINT : PRINT "DATA FOR GRAPH": IF K$ = "Y" THEN
       PRINT "STORED AS FILE ";G$: PRINT
1150   IF HR <  > 0 THEN 1243
1160   PRINT : PRINT "ORIGIN AT ZERO TIME AND Y = ";Y0
       : PRINT
1170   PRINT "TIME INTERVAL IS ";T(W);" SECS ": PRINT

1180   GOTO 1245
1243   PRINT : PRINT "ORIGIN AT T = ";K0;" AND Y = ";Y
       0: PRINT
1244   PRINT "T RANGE IS ";K0;" K TO ";KF;" K ": PRINT

1245   PRINT "Y RANGE IS ";Y0;" TO ";Y9;" ON 12 BIT SC
       ALE FROM 0 TO 4095"
1246   IF HR = 0 THEN  PRINT "TIC INTERVAL ON TIME AXI
       S IS ";T(W) / 10;" SECS ": GOTO 1248
1247   PRINT "TIC INTERVAL ON T AXIS IS ";D1 / 10;" K
       ": PRINT "TIC INTERVAL ON Y AXIS IS ";D2 / 10
1248   PRINT
1249   PRINT "HEATING RATE WAS ";HR;" K/MIN": PRINT
1250   PRINT : PRINT "ACTUAL MAX Y = ";MAX
1300   PRINT "ACTUAL MIN Y = ";B: PRINT
1400   PR# 0
1580   GOTO 2000
1585   REM ----------------WRITE TO DISK-----------
1590   PRINT : FLASH : PRINT "PLEASE WAIT - WRITING TO
       DISK ": NORMAL
1600   PRINT D$;"OPEN";G$
1610   PRINT D$;"WRITE";G$
1620   FOR I = 0 TO W
1622   IF HR = 0 THEN 1628
1624   PRINT R(I)
1626   GOTO 1630
1628   PRINT T(I)
1630   PRINT V(I)
1640   NEXT I
1645   PRINT 9999: PRINT 9999
1650   PRINT D$;"CLOSE"
1660   GOTO 659
1990   PRINT
2000   FLASH : PRINT "RUN ENDED ": NORMAL
2010   END
```

```
6999   REM --------------DUMP SUBROUTINE-----------
7000   PRINT "IT IS POSSIBLE TO GET A PRINTOUT OF YOUR
       GRAPH "
7010   FOR I = 1 TO 3000: NEXT I
7020   PRINT : PRINT "DO YOU WANT TO ATTEMPT A PRINTOU
       T ? Y OR N ?"
7030   INPUT Z$: IF Z$ < > "Y" THEN 7210
7040   PRINT : INVERSE : INPUT "ENTER PRINTER TYPE : E
       PSON = E , C-ITOH = C ";Y$
7050   NORMAL
7060   IF Y$ < > "E" AND Y$ < > "C" THEN 7040
7070   IF Y$ = "E" THEN 7190
7080   D$ = CHR$ (4):E$ = CHR$ (27)
7090   INVERSE : PRINT "REMEMBER APPLE PARALLEL INTERF
       ACE IN SLOT 1 ": NORMAL
7100   FOR I = 1 TO 1000: NEXT I
7110   PRINT D$;"PR#1": PRINT E$;"P": REM  PROPORTIONA
       L PRINT
7120   PRINT D$;"PR#0"
7130   PRINT D$;"BLOAD CITOH.OBJ"
7140   CALL 6400
7150   PRINT D$;"PR#1"
7160   PRINT E$;"N": REM  RESET TO PICA
7170   PRINT D$;"PR#0"
7180   GOTO 7210
7190   PR# 1
7200   PRINT  CHR$ (17)
7210   RETURN
```

B.3 Display of data

Although the program DSC provides for display of the data captured and stored, it is often convenient to have a separate program for displaying the contents of the data files which soon accumulate. Decisions can then be taken on further procedures to be followed.

Program FASTDISP: This is a general-purpose program for displaying the contents X, Y of an APPLE sequential file.

> Input: Name of the file containing the data.
> Output: Display on screen of a plot of X from XMIN to XMAX against Y on a scale extended 10% above and below the YMIN and YMAX values. The points are joined (see line 1080) and the display is automatically dumped to an Epson printer.

Program READISP: This is a more versatile, but slower, version of FASTDISP and includes option for deletion and addition of points.

```
1   REM      CHECK THAT PROGRAM IS LOADED ABOVE SCREEN#2

2   IF  PEEK (104) = 96 THEN 10
3   POKE 103,1: POKE 104,96: POKE 4096 * 6,0
4   PRINT  CHR$ (4)"RUN FASTDISP"
10   HOME
12   PRINT "*****************************************"
13   PRINT "*                                       *"
14   PRINT "* PROGRAM 'FASTDISP'                    *"
15   PRINT "*FOR QUICK DISPLAY OF DATA FROM A FILE*"
18   PRINT "*                                       *"
19   PRINT "*****************************************"
30   PRINT : PRINT "VERSION AS AT  1/08/86": PRINT
50   PRINT : PRINT "N.B. PRINTER MUST BE AN EPSON AND
     MUST BE ON ": PRINT : PRINT "POINTS WILL BE JOINE
     D SEE LINE 1080"
60   D$ =  CHR$ (4)
70   DIM A(502),B(502)
80   PRINT : INPUT "ENTER FILENAME ";G$
90   PRINT D$;"OPEN";G$
100   PRINT D$;"READ";G$
105   PRINT : INVERSE : PRINT "PLEASE WAIT - READING":
      NORMAL : PRINT
110  I = 1
120   INPUT A(I),B(I)
140   IF A(I) = 9999 GOTO 170
150  I = I + 1
160   GOTO 120
170   PRINT D$;"CLOSE"
180   PRINT "END OF FILE"
190   PRINT : PRINT "OUTPUT TO PRINTER NEXT"
660   XMAX = A(1):MX = A(1)
670   YMAX = B(1):MY = B(1)
680   FOR J = 2 TO I
690   IF A(J) = 9999 GOTO 746
700   IF A(J) > XMAX THEN XMAX = A(J)
710   IF A(J) < MX THEN MX = A(J)
720   IF B(J) > YMAX THEN YMAX = B(J)
730   IF B(J) < MY THEN MY = B(J)
740   NEXT J
746   PR# 1
749   PRINT : PRINT "FILENAME = ";G$
750   PRINT : PRINT "XMAX = ",XMAX
760   PRINT "XMIN = ",MX
770   PRINT "YMAX = ",YMAX
780   PRINT "YMIN = ",MY
785   PR# 0
1030 X9 = XMAX:XO = MX
1035 RANGE = YMAX - MY
1040 Y9 = YMAX + 0.05 * RANGE:YO = MY - 0.05 * RANGE
1050 W = I: REM  NO. OF POINTS
```

```
1060 D1 = X9 - X0
1070 D2 = Y9 - Y0
1080 J$ = "Y": REM  POINTS JOINED
1090   NORMAL
1100   REM  -OUTLINE FRAME
1110   HGR
1120   HCOLOR= 3
1130   HPLOT 9,0 TO 279,0 TO 279,150 TO 9,150 TO 9,0
1140   REM   TIC INTERVALS
1142 G4 =  INT (270 / 10)
1144 G5 =  INT (150 / 10)
1146   FOR K = 1 TO 9
1148   HPLOT 9 + K * G4,150 TO 9 + K * G4,147
1150   HPLOT 9,147 - K * G5 TO 12,147 - K * G5
1152   NEXT K
1160   FOR I = 1 TO W
1180 A = A(I)
1190 B = B(I)
1210   REM -PLOTTING OF THE POINTS
1220 X =  INT (270 * (A - X0) / D1) + 9
1230 Y = 150 -  INT (150 * (B - Y0) / D2)
1240   IF X < 0 OR X > 279 OR Y < 0 OR Y > 150 THEN 12
     80
1250   IF I = 1 OR V = 0 OR U = 0 OR J$ <  > "Y" THEN
     HPLOT X,Y: GOTO 1270
1260   HPLOT V,U TO X,Y
1270 V = X:U = Y
1280   NEXT I
1300   PRINT
1540   PR# 1
1550   PRINT  CHR$ (17)
1560   PR# 0
1570   PR# 1
1580   PRINT : PRINT "DATA FOR GRAPH STORED AS ";G$
1590   PRINT : PRINT "ORIGIN AT X = ";X0;" AND Y = ";Y
     0
1600   PRINT : PRINT "X RANGE IS ";X0;" TO ";X9
1610   PRINT : PRINT "Y RANGE IS ";Y0;" TO ";Y9
1612   PRINT : PRINT "X TIC INTERVAL = ";D1 / 10
1614   PRINT : PRINT "Y TIC INTERVAL = ";D2 / 10
1620   PR# 0
1630   END
```

```
1   REM      CHECK THAT PROGRAM IS LOADED ABOVE SCREEN#2

2   IF  PEEK (104) = 96 THEN 10
3   POKE 103,1: POKE 104,96: POKE 4096 * 6,0
4   PRINT  CHR$ (4)"RUN READISP"
10   HOME
12   PRINT "****************************************"
13   PRINT "*       PROGRAM 'READISP'              *"
14   PRINT "*  TO READ X,Y DATA FROM A TEXTFILE    *"
15   PRINT "*                                      *"
16   PRINT "****************************************"
19   PRINT : PRINT
20   PRINT "VERSION AS AT 1/08/86": PRINT
50   PRINT : PRINT "POINTS MAY BE DELETED OR ADDED ": PRINT

60  D$ =  CHR$ (4)
70   DIM A(352),B(352)
80   PRINT : INPUT "ENTER FILENAME ";G$
90   PRINT D$;"OPEN";G$
100   PRINT D$;"READ";G$
110  I = 1
120   INPUT A(I),B(I)
130   PRINT A(I),B(I)
140   IF A(I) = 9999 GOTO 170
150  I = I + 1
160   GOTO 120
170   PRINT D$;"CLOSE"
180   PRINT "END OF FILE"
181   PRINT : INPUT "LIST CONTENTS OF FILE ON PRINTER
     ? Y OR N ? ";Z$
182   IF Z$ <  > "Y" THEN 190
183   PR# 1
184   PRINT : PRINT : PRINT "CONTENTS OF FILE ";G$: PRINT

185   PRINT "A(I)","B(I)": PRINT
186   FOR N = 1 TO I
187   PRINT A(N),B(N)
188   NEXT N
189   PR# 0
190   PRINT : INPUT "DO YOU WANT TO DELETE ANY POINTS
     ? ";F$
200   IF F$ <  > "Y" GOTO 360
202   PRINT : INPUT "BY POINT NUMBER OR VALUE ? P OR V
     ? ";F$
204   IF F$ <  > "P" THEN 210
206   PRINT : INPUT "ENTER POINT NO.(REMEMBER THAT EAC
     H DELETION IS FOLLOWED BY REORDERING) > ";K
207   IF K > I OR K < = 0 THEN 206
208   GOTO 280
210   PRINT : PRINT "ENTER THE POINT TO BE DELETED : X
     ,Y "
220   INPUT X,Y
230   FOR J = 1 TO I
```

```
240   IF A(J) = X AND B(J) = Y THEN 270
250   NEXT J
260   PRINT : PRINT "POINT NOT FOUND"
265   GOTO 190
270 K = J
280   FOR J = K + 1 TO I
290 A(J - 1) = A(J)
300 B(J - 1) = B(J)
310   NEXT J
320 I = I - 1
330   PRINT : INPUT "DO YOU WANT TO DELETE ANY MORE PO
      INTS ? ";H$
340   IF H$ = "N" GOTO 360
350   GOTO 202
360   PRINT : INPUT "DO YOU WANT TO ADD ANY POINTS ? "
      ;L$
370   IF L$ = "N" GOTO 560
380   PRINT : PRINT "ENTER THE POINT TO BE ADDED : X,Y
      "
390   INPUT X,Y
400   FOR J = 1 TO I
410   IF A(J) > X GOTO 430
420   NEXT J
430 K = J
440 J = I
450   IF J = K - 1 GOTO 500
460 A(J + 1) = A(J)
470 B(J + 1) = B(J)
480 J = J - 1
490   GOTO 450
500 I = I + 1
510 A(K) = X
520 B(K) = Y
530   PRINT : INPUT "DO YOU WANT TO ADD ANY MORE POINT
      S ? ";M$
540   IF M$ = "N" GOTO 560
550   GOTO 380
560   IF F$ = "N" AND L$ = "N" GOTO 640
570   PRINT D$;"OPEN";G$
580   PRINT D$;"WRITE";G$
590   FOR J = 1 TO I
600   PRINT A(J): PRINT B(J)
610   NEXT J
620   PRINT D$;"CLOSE";G$
630   PRINT : PRINT "CORRECTED DATA STORED IN FILE ";G
      $
640   PRINT : INPUT "DO YOU WANT A PRINTOUT OF MAX AND
      MIN   VALUES ? Y OR N ? ";C$
650   IF C$ = "N" GOTO 790
660 XMAX = A(1):MX = A(1)
670 YMAX = B(1):MY = B(1)
680   FOR J = 2 TO I
```

```
690    IF A(J) = 9999 GOTO 745
700    IF A(J) > XMAX THEN XMAX = A(J)
710    IF A(J) < MX THEN MX = A(J)
720    IF B(J) > YMAX THEN YMAX = B(J)
730    IF B(J) < MY THEN MY = B(J)
740    NEXT J
745    IF Z$ = "N" THEN 750
746    PR# 1
750    PRINT : PRINT "XMAX = ",XMAX
760    PRINT "XMIN = ",MX
770    PRINT "YMAX = ",YMAX
780    PRINT "YMIN = ",MY
785    PR# 0
790    REM   END OF ORIGINAL"READFILE"
800    REM   START OF ORIGINAL"FILEDISPLAY"
1000   PRINT : INPUT "DO YOU WANT TO DISPLAY PLOT  ON
       SCREEN ? Y OR  N ? ";Z$
1010   IF Z$ <  > "Y" THEN 1340
1020 M$ = G$: REM  FILENAME
1030 X9 = XMAX:XO = MX
1035 RANGE = YMAX - MY
1040 Y9 = YMAX + 0.05 * RANGE:YO = MY - 0.05 * RANGE
1050 W = I: REM  NO. OF POINTS
1060 D1 = X9 - XO
1070 D2 = Y9 - YO
1080   PRINT : INPUT "DO YOU WANT POINTS JOINED ? Y OR
       N ? ";J$
1090   NORMAL
1100   REM   -OUTLINE FRAME
1110   HGR
1120   HCOLOR= 3
1130   HPLOT 9,0 TO 279,0 TO 279,150 TO 9,150 TO 9,0
1140 D$ =  CHR$ (4)
1150   PRINT D$;"OPEN";M$
1160   FOR I = 1 TO W
1170   PRINT D$;"READ";M$
1180   INPUT A
1190   INPUT B
1200   IF A = 9999 THEN 1320
1210   REM -PLOTTING OF THE POINTS
1220 X =   INT (270 * (A - XO) / D1) + 9
1230 Y = 150 -   INT (150 * (B - YO) / D2)
1240   IF X < 0 OR X > 279 OR Y < 0 OR Y > 150 THEN 12
       80
1250   IF I = 1 OR V = 0 OR U = 0 OR J$ <  > "Y" THEN
       HPLOT X,Y: GOTO 1270
1260   HPLOT V,U TO X,Y
1270 V = X:U = Y
1280   NEXT I
1290   PRINT D$;"CLOSE";M$
1300   PRINT
1310   PRINT : GOTO 1330
```

```
1320   PRINT : PRINT "END OF FILE REACHED ": GOTO 1290

1330   GOSUB 1350
1340   END
1350   PRINT "IT IS POSSIBLE TO GET A PRINTOUT OF YOUR
       GRAPH "
1360   FOR I = 1 TO 3000: NEXT I
1370   PRINT : PRINT "DO YOU WANT TO ATTEMPT A PRINTOU
       T ? Y OR N ?"
1380   INPUT Z$: IF Z$ < > "Y" THEN 1570
1390   PRINT : INVERSE : INPUT "ENTER PRINTER TYPE : E
       PSON = E , C-ITOH = C ";Y$
1400   NORMAL
1410   IF Y$ < > "E" AND Y$ < > "C" THEN 1390
1420   IF Y$ = "E" THEN 1540
1430 D$ = CHR$ (4):E$ = CHR$ (27)
1440   INVERSE : PRINT "REMEMBER APPLE PARALLEL INTERF
       ACE IN SLOT 1 ": NORMAL
1450   FOR I = 1 TO 1000: NEXT I
1460   PRINT D$;"PR#1": PRINT E$;"P": REM  PROPORTIONA
       L PRINT
1470   PRINT D$;"PR#0"
1480   PRINT D$;"BLOAD CITOH.OBJ"
1490   CALL 6400
1500   PRINT D$;"PR#1"
1510   PRINT E$;"N": REM  RESET TO PICA
1520   PRINT D$;"PR#0"
1530   GOTO 1570.
1540   PR# 1
1550   PRINT  CHR$ (17)
1560   PR# 0
1570   PR# 1
1580   PRINT : PRINT "DATA FOR GRAPH STORED AS ";M$
1590   PRINT : PRINT "ORIGIN AT X = ";X0;" AND Y = ";Y
       0
1600   PRINT : PRINT "X RANGE IS ";X0;" TO ";X9
1610   PRINT : PRINT "Y RANGE IS ";Y0;" TO ";Y9
1620   PR# 0
1630   RETURN
```

B.4 Modification of the data

(a) Baseline correction

Program OPERFILE: When the baseline for a run is not flat, it is advisable to record the baseline separately, using program DSC. The baseline can then be subtracted from the sample data using program OPERFILE. The program is more general in that it allows addition, subtraction, multiplication or division of the contents of two files.

Input: Names of two files, choice of operation and a name of a file for output.

Output: Results of the operation stored in the file specified. Summary of operations on the printer.

```
10    HOME
20    PRINT "*****************************************"
30    PRINT "* PROGRAM 'OPERFILE' TO ADD,SUBTRACT,  *"
40    PRINT "* MULTIPLY OR DIVIDE THE Y VALUES FROM *"
50    PRINT "* TWO TEXTFILES                        *"
55    PRINT "*****************************************"
56    PRINT : PRINT "VERSION AS AT 28/06/85   MEB": PRINT

57    PRINT "FILES MAY BE EDITED USING 'READFILE'": PRINT
      "            SMOOTHED USING 'SMOOTHFILE'": PRINT
      "          OR DIFFERENTIATED USING 'NUMDIFILE'": FLASH
      : PRINT "              BEFOREHAND             ": NORMAL
      : PRINT : PRINT
60    D$ =  CHR$ (4)
65    PRINT : INPUT "IS PRINTER ON ? Y OR N ? ";Z$
66    IF Z$ <  > "Y" THEN 65
70    DIM A(502),B(502)
75    DIM C(500),D(500),AA(500),F(500)
80    PRINT : INPUT "ENTER FIRST FILENAME, NOTE ORDER O
      F OPERATION IS FIRST+,-,*,/SECOND   > ";G$
90    PRINT D$;"OPEN";G$
100   PRINT D$;"READ";G$
110 I = 1
120   INPUT A(I),B(I)
130   PRINT A(I),B(I)
140   IF A(I) = 9999 GOTO 165
150 I = I + 1
160   GOTO 120
165 GM = I - 1
170   PRINT D$;"CLOSE"
180   PRINT "END OF FILE"
190   PRINT : INPUT "ENTER SECOND FILENAME    >   ";H$
195   PRINT D$;"OPEN";H$
200   PRINT D$;"READ";H$
210 I = 1
220   INPUT C(I),D(I)
230   PRINT C(I),D(I)
240   IF C(I) = 9999 GOTO 265
250 I = I + 1
260   GOTO 220
265 HM = I - 1
270   PRINT D$;"CLOSE"
280   PRINT "END OF FILE"
```

```
290    PR# 1
300    POKE 1657,80
310    PRINT : PRINT "FILE NO.1 = ";G$
320    PRINT : PRINT "FIRST POINT IS ";A(1); SPC( 4);B(
       1)
330    PRINT : PRINT "LAST POINT IS  ";A(GM); SPC( 4);B
       (GM)
335    PRINT : PRINT "NO. OF POINTS IN FILE,EXCLUDING 9
       999,0 WAS ";GM
340    PRINT : PRINT : PRINT
350    PRINT : PRINT "FILE NO.2 = ";H$
360    PRINT : PRINT "FIRST POINT IS ";C(1); SPC( 4);D(
       1)
370    PRINT : PRINT "LAST POINT IS  ";C(GM); SPC( 4);D
       (GM)
375    PRINT : PRINT "NO. OF POINTS IN FILE,EXCLUDING 9
       999,0 WAS ";HM
380    PRINT : PRINT : PRINT
400    PR# 0
410    PRINT : PRINT "CHOOSE THE OPERATION REQUIRED": PRINT
       "(1)   ADDITION ": PRINT "(2)   SUBTRACTION ": PRINT
       "(3)   MULTIPLICATION ": PRINT "(4)   DIVISION ": INPUT
       CH
420    IF CH < 1 OR CH > 4 THEN 410
450    PRINT : INPUT "ENTER NUMBER OF STARTING POINT IN
       FILE 1  > ";G1
460    PRINT : INPUT "ENTER NUMBER OF STARTING POINT IN
       FILE 2  > ";H1
465 GO = G1:HO = H1
470    PRINT : INPUT "ENTER NO. OF POINTS TO BE TREATED
        > ";NP
490    FOR I = 1 TO NP
500    ON CH GOTO 510,520,530
505    GOTO 540
510 AA(I) = B(G1) + D(H1): REM      ADDITION
515    GOTO 550
520 AA(I) = B(G1) - D(H1): REM      SUBTRACTION
525    GOTO 550
530 AA(I) = B(G1) * D(H1): REM      MULTIPLICATION
535    GOTO 550
540 AA(I) = B(G1) / D(H1): REM      DIVISION
550 K = GO + I - 1
552 F(I) = A(K)
554 G1 = G1 + 1:H1 = H1 + 1
555    PRINT F(I); SPC( 4);AA(I)
560    NEXT I
562    PRINT : INPUT "DO YOU WANT TO STORE OUTPUT IN A
       FILE? Y OR N ?";C$
563    IF C$ = "N" THEN 640
565    PRINT : INPUT "ENTER A FILENAME FOR STORAGE  > "
       ;L$
570    PRINT D$;"OPEN";L$
```

```
580   PRINT D$;"WRITE";L$
590   FOR I = 1 TO NP
600   PRINT F(I): PRINT AA(I)
610   NEXT I
615   PRINT 9999: PRINT 0
620   PRINT D$;"CLOSE";L$
630   PRINT : PRINT "DATA STORED IN FILE ";L$
640   PRINT : INPUT "DO YOU WANT A PRINTOUT OF MAX AND
      MIN   VALUES ? Y OR N ? ";C$
650   IF C$ = "N" GOTO 790
660   XMAX = F(1):MX = F(1)
670   YMAX = AA(1):MY = AA(1)
680   FOR J = 2 TO NP
690   IF F(J) = 9999 GOTO 750
700   IF F(J) > XMAX THEN XMAX = F(J)
710   IF F(J) < MX THEN MX = F(J)
720   IF AA(J) > YMAX THEN YMAX = AA(J)
730   IF AA(J) < MY THEN MY = AA(J)
740   NEXT J
750   PRINT : PRINT "XMAX = ",XMAX
760   PRINT "XMIN = ",MX
770   PRINT "YMAX = ",YMAX
780   PRINT "YMIN = ",MY
790   PRINT : INPUT "DO YOU WANT A SUMMARY ON THE PRIN
      TER ? N.B. SWITCH ON ! Y OR N ? ";C$
800   IF C$ <  > "Y" THEN 1000
802   PRINT : INPUT " FULL=F, SHORT=S  > ";Q$
810   PR# 1
820   POKE 1657,80
830   PRINT : PRINT : PRINT "SUMMARY OF OPERATION ON F
      ILES ";G$;" AND ";H$
840   PRINT "========================================
      ========": PRINT
850   PRINT "THE FILES WERE ";
860   ON CH GOTO 880,890,900
870   GOTO 910
880   PRINT "ADDED ": PRINT : GOTO 920
890   PRINT "SUBTRACTED ": PRINT : GOTO 920
900   PRINT "MULTIPLIED ": PRINT : GOTO 920
910   PRINT "DIVIDED ": PRINT
920   PRINT "ORDER WAS FILE ";G$;" +,-,* OR / ";" FILE
      ";H$: PRINT
930   PRINT "STARTING POINTS WERE POINT ";G0;" IN FILE
      ";G$;" AND POINT ";H0;" IN FILE ";H$: PRINT : PRINT
      NP;" POINTS WERE TREATED ": PRINT
940   PRINT : PRINT
945   IF Q$ = "S" THEN 990
950   FOR I = 1 TO NP
960   PRINT F(I); SPC( 4);AA(I)
970   NEXT I
990   PRINT : PRINT "XMAX = ",XMAX
991   PRINT "XMIN = ",MX
```

```
992   PRINT "YMAX = ",YMAX
993   PRINT "YMIN = ",MY
994   IF L$ = "       " THEN 999
995   PRINT : PRINT "DATA STORED IN FILE ";L$: PRINT
999   PR# 0
1000  END
```

(b) Smoothing of data

There are many approaches to this problem (chapter 12 and Fig. 12.3). Three programs are given below for smoothing of noisy data, using different algorithms. The algorithms are given in the references quoted in the listings.

Programs SMOOTHFILE, BINARY FILTER, ROBUST FILTER

   Input: Filenames and adjustable parameters as specified.
   Output: Smoothed data in a file, summary of operations on the printer.

```
10    HOME
20    PRINT "******************************************"
30    PRINT "* DATA SMOOTHING PROGRAM                 *"
40    PRINT "* BASED ON SAVITZKY GOLAY FILTER         *"
50    PRINT "* REF:ANAL.CHEM.,36(1964)1627            *"
60    PRINT "* USING 5 POINT CONVOLUTION TABLE I      *"
70    PRINT "******************************************"
80    PRINT : PRINT
90    D$ =  CHR$ (4)
100   DIM Y(250),Z(250),D(5),C(5)
110   I = 0: REM   COUNTER
120   PRINT "DATA MUST BE AT EQUALLY-SPACED INTERVALS"

130   PRINT
140   PRINT : PRINT "DATA MUST END WITH 9999,0 "
150   PRINT : INPUT "ENTER TIME INTERVAL IN MINUTES ";
      X
160   PRINT : INPUT "ENTER NAME OF DATAFILE ";G$
170   PRINT D$;"OPEN";G$
180   PRINT D$;"READ";G$
190   FOR I = 1 TO 350
200   INPUT T
210   INPUT Y(I)
220   IF T = 9999 THEN 240
230   NEXT I
240   PRINT D$;"CLOSE";G$
250   PRINT : PRINT "    CALCULATING  -  PLEASE WAIT"
260   M = I - 1: REM   NO OF POINTS ENTERED
270   REM - CONVOLUTING FACTORS FOR 5 POINT SMOOTHING
280   C(1) =   - 3:C(2) = 12:C(3) = 17:C(4) = 12:C(5) =
      - 3
```

```
290 N = 35: REM - NORMALISING FACTOR
300  REM - READ 5 Y VALUES INTO TEMPORARY STORE
310 K = 0: REM - POINT COUNTER
320 S = 0: REM - SUMMATION
330  REM - MULTIPLY Y VALUES BY CONVOLUTING FACTORS,S
    UM AND THEN NORMALISE TO GET NEW VALUE OF THE MID
    -POINT
340  FOR J = 1 TO 5
350 D(J) = Y(K + J)
360 D(J) = C(J) * D(J)
370 S = S + D(J)
380  NEXT J
390  IF K > 5 THEN 430
400  REM - STORE FIRST TWO POINTS
410 Z(1) = Y(1)
420 Z(2) = Y(2)
430 Z(K + 3) = S / N
440 K = K + 1
450  IF K <  = (M - 5) THEN 320
460 Z(M - 1) = Y(M - 1)
470 Z(M) = Y(M)
480  REM - LAST TWO POINTS STORED
490  PRINT : INPUT "ENTER FILENAME FOR SMOOTHED VALUE
    S ";F$
500  PRINT D$;"OPEN";F$
510  PRINT D$;"WRITE";F$
520  FOR L = 1 TO M
530  PRINT L * X,Z(L)
540  NEXT L
550  PRINT "9999": PRINT "0"
560  PRINT D$;"CLOSE";F$
570  PRINT : INPUT "DO YOU WANT MORE SMOOTHING ? Y OR
    N ? ";B$
580  IF B$ = "Y" THEN 600
590  END
600  FOR L = 1 TO M
610 Y(L) = Z(L)
620  NEXT L
630  REM - SMOOTHED VALUES NOW BECOME INPUT
640  GOTO 310
```

```
10  HOME
20  PRINT "****************************************"
30  PRINT "* DATA SMOOTHING PROGRAM              *"
40  PRINT "* BASED ON BINOMIAL FILTER            *"
50  PRINT "* REF: REV.SCI.INST.,54(8)(1984)1034  *"
60  PRINT "*      MARCHAND AND MARMET            *"
70  PRINT "****************************************"
```

```
75   PRINT : PRINT "VERSION 7/5/86"
80   PRINT : PRINT
90 D$ =  CHR$ (4)
100  DIM Y(500),Z(500),X(500)
110 I = 0: REM  COUNTER
115 J = 0: REM  PASS COUNTER
120  PRINT "DATA MUST BE AT APPROXIMATELY EQUALLY-SPA
     CED INTERVALS"
130  PRINT
140  PRINT : PRINT "DATA MUST END WITH 9999,0 "
160  PRINT : INPUT "ENTER NAME OF DATAFILE ";G$
170  PRINT D$;"OPEN";G$
180  PRINT D$;"READ";G$
190  FOR I = 1 TO 500
200  INPUT X(I)
210  INPUT Y(I)
220  IF X(I) = 9999 THEN 240
230  NEXT I
240  PRINT D$;"CLOSE";G$
245  PRINT : INPUT "SET NO. OF PASSES EG. 25 ";N
250  PRINT : PRINT "   CALCULATING  -  PLEASE WAIT"
260 M = I - 1: REM  NO OF POINTS ENTERED
265 J = J + 1
270  FOR I = 1 TO (M - 1)
280 Z(I) = (Y(I) + Y(I + 1)) / 2
290  NEXT I
300  FOR I = 2 TO (M - 1)
310 Y(I) = (Z(I - 1) + Z(I)) / 2
320  NEXT I
330  IF J <  = N THEN 265
380  PRINT : INPUT "OUTPUT ON SCREEN (S)  , PRINTER (P
     ) OR ONLY TO DISK (D) ? ";Z$
390  IF Z$ = "D" THEN 490
395  IF Z$ = "S" THEN 400
398  PR# 1
400  FOR L = 1 TO M
410  PRINT X(L),Y(L)
420  NEXT L
425  PR# 0
490  PRINT : INPUT "ENTER FILENAME FOR SMOOTHED VALUE
     S ";F$
500  PRINT D$;"OPEN";F$
510  PRINT D$;"WRITE";F$
520  FOR L = 1 TO M
530  PRINT X(L): PRINT Y(L)
540  NEXT L
550  PRINT "9999": PRINT "0"
560  PRINT D$;"CLOSE";F$
590  END
```

```
10   HOME
20   PRINT "*************************************": PRINT

30   PRINT "* PROGRAM 'ROBUST FILTER'          *"
31   PRINT : PRINT "**********************************
     **"
35   PRINT : PRINT "VERSION AS AT   3/07/85 MWB/MEB "
41   PRINT : PRINT "BASED ON B-M.BUSSIAN AND W.HARDLE,
     J.APPL.SPECTROSC.,38(1984)309-313 ": PRINT
42   PRINT "SMOOTHING IS TUNED BY TWO PARAMETERS ": PRINT
     "1.SIGNAL-TO-NOISE RATIO": PRINT "2. HALF-WIDTH O
     F DATA PEAKS "
43   PRINT "THESE PARAMETERS HAVE TO BE SET IN LINE 50
      (AS S AND P ) ACCORDING TO THE SCALE OF THE INPU
     T DATA AND BY TRIAL AND ERROR "
50   S = 1:P = 50
51   INVERSE
52   PRINT "PRESENT SETTING : S = ";S;" P = ";P
53   NORMAL
55   PRINT : PRINT "A ' WINDOW SIZE ' IS NOW SET ": PRINT

60   REM      - WINDOW DIMENSION
70   INPUT "INPUT WINDOW HEIGHT (E.G. 50)";K
80   INPUT "INPUT WINDOW WIDTH (ODD NO. 3-9 )";E
90   DIM X(500),Y(501),W(23),Z(23)
100  PRINT : INPUT "ENTER FILENAME FROM WHICH DATA IS
     TO BE READ  > ";A$
110  PRINT : INPUT "ENTER FILENAME  IN  WHICH DATA IS
     TO BE STORED  > ";B$
120  PRINT : FLASH : PRINT "READING DATA FROM DISK !
     ": NORMAL
130  D$ =  CHR$ (4)
140  PRINT D$;"OPEN";A$
150  PRINT D$;"READ";A$
160  FOR I = 1 TO 500
170  INPUT X(I): INPUT Y(I)
180  IF X(I) = 9999 THEN 200
190  NEXT I
200  G = I - 1
210  PRINT D$;"CLOSE";A$
220  HOME
230  GOSUB 450
240  FOR N =  INT (E / 2) TO G -  INT (E / 2)
250  GOSUB 550
260  Y(N) =  INT (T)
270  PRINT X(N),Y(N)
280  NEXT N
290  PRINT : FLASH : PRINT "OPENING FILE ";B$: NORMAL
300  PRINT D$;"OPEN";B$
310  PRINT D$;"WRITE";B$
320  FOR I = 1 TO N - 1
330  PRINT X(I): PRINT Y(I)
```

```
340    NEXT I
350    PRINT 9999: PRINT 9999
360    PRINT D$;"CLOSE";B$
370    PRINT : PRINT "DO YOU WANT A SUMMARY ON PRINTER
       ? Y OR N ? ": INPUT Z$
380    IF Z$ <  > "Y" THEN 410
390    PR# 1
400    POKE 1657,80
410    PRINT : PRINT "SUMMARY OF OPERATIONS": PRINT "==
       ========================": PRINT
420    PRINT I - 1;" DATA POINTS FROM FILE ";A$: PRINT
       "SMOOTHED USING ROBUST FILTER": PRINT "OUTPUT STO
       RED IN FILE ";B$
425    PRINT : PRINT "WINDOW HEIGHT SET AT ";K
426    PRINT "WINDOW WIDTH SET AT ";E
427    PRINT : PRINT
430    PR# 0
440    END
450    REM
460    REM   - WINDOW SUBROUTINE
470 M =   INT (E / 2 + 1)
480    FOR I = 1 TO M - 1
490 X = S * I / M
500 W = 0.75 * (1 - X * X)
510 W(M + I) = W:W(M - I) = W
520    NEXT I
530 W(M) = 0.75
540    RETURN
550    REM
560    REM   - SMOOTHING SUBROUTINE
570 L = N -   INT (E / 2)
580    FOR I = 1 TO E
590 Z(I) = Y(L + I - 1)
600    NEXT I
610 T = Y(N):D = 0
620 S1 = 0:S2 = 0
630    FOR I = 1 TO E
640 Y1 = Z(I) - T
650 W = W(I)
660 Y2 = K:Y3 = 0
670    IF Y1 > K THEN 710
680 Y2 =   - K
690    IF Y1 <   - K THEN 710
700 Y2 = Y1:Y3 = 1
710 S1 = S1 + (W * Y2)
720 S2 = S2 + (W * Y3)
730    NEXT I
740    IF S2 = 0 THEN S2 = 0.75
750 H = S1 / S2
760    IF ( ABS (H) <  = P) THEN  RETURN
770 T = T + H
780    GOTO 620
```

(c) Scaling of data

It is sometimes useful, e.g., in kinetic analyses, to be able to scale either the X or the Y data, or both.

Program SCALEFILE

Input: The filename, the X and Y scale factors and the Y base correction or offset. A filename for storage of the scaled data.

Output: Scaled data in a file.

```
10   HOME
20   PRINT "****************************************"
30   PRINT "* PROGRAM 'SCALEFILE' TO MULTIPLY      *"
40   PRINT "* THE X AND Y VALUES FROM A TEXTFILE   *"
50   PRINT "* BY    CONSTANT FACTORS               *"
55   PRINT "****************************************"
56   PRINT : PRINT "VERSION AS AT19/07/85  MEB": PRINT

60 D$ =  CHR$ (4)
70   DIM A(500),B(500)
75   DIM C(500),D(500),AA(500),F(500)
80   PRINT : INPUT "ENTER FILENAME > ";G$
85   PRINT : INPUT "ENTER X SCALE FACTOR > ";XF
86   PRINT : INPUT "ENTER Y BASE CORRECTION   FACTOR >
     ";YB
87   PRINT : INPUT "ENTER Y SCALE FACTOR > ";YF
89   PRINT : INVERSE : PRINT "PLEASE WAIT , READING ":
     NORMAL : PRINT
90   PRINT D$;"OPEN";G$
100  PRINT D$;"READ";G$
110  I = 1
120  INPUT A(I),B(I)
121  IF A(I) = 9999 GOTO 165
129  A(I) = A(I) * XF
130  B(I) = (B(I) - YB) * YF
150  I = I + 1
160  GOTO 120
165 GM = I - 1
170  PRINT D$;"CLOSE"
180  PRINT "END OF FILE"
565  PRINT : INPUT "ENTER A FILENAME FOR STORAGE > "
     ;L$
569  PRINT : INVERSE : PRINT "PLEASE WAIT , WRITING "
     : NORMAL : PRINT
570  PRINT D$;"OPEN";L$
580  PRINT D$;"WRITE";L$
590  FOR I = 1 TO GM
600  PRINT A(I): PRINT B(I)
610  NEXT I
615  PRINT 9999: PRINT 0
```

```
620   PRINT D$;"CLOSE";L$
630   PRINT : PRINT "DATA STORED IN FILE ";L$
1000  END
```

B.5 Processing of the data

(a) Numerical differentiation

The DSC curve is already an example of a differential measurement and so is not usually differentiated further. If an integral measurement such as a TG curve has been recorded, then a DTG curve can be obtained by numerical differentiation. The usual method is that of Savitsky and Golay (reference given in the listing of program NUMDIFILE).

Program NUMDIFILE

> Input:　Option of entering data manually or from a file. The X data must be equally spaced. Option for preliminary smoothing as in program SMOOTHFILE, and then differentiation may be based on either (1) a quadratic or (2) a cubic polynomial.
>
> Output: Tables of values on printer and storage of values in a file.

```
10    HOME
20    PRINT "****************************************"
30    PRINT "* NUMERICAL DIFFERENTIATION PROGRAM    *"
40    PRINT "* BASED ON SAVITZKY AND GOLAY          *"
50    PRINT "* REF:ANAL.CHEM.;36(1964)1627          *"
60    PRINT "* USING 5 POINT CONVOLUTION TABLE      *"
70    PRINT "****************************************"
80    PRINT : PRINT
90    DIM Y(250),Z(250),D(5),C(5)
100   PRINT "YOU MAY INPUT DATA MANUALLY(M) "
110   PRINT "OR FROM A DATA FILE(F) "
120   INPUT "ENTER M OR F ";Q$
130   IF Q$ < > "F" THEN 340
140   PRINT : INPUT "ENTER FILENAME ";P$
150 D$ = CHR$ (4)
160   PRINT D$;"OPEN";P$
170   PRINT D$;"READ";P$
180   FOR I = 1 TO 3
190   INPUT X(I)
200   INPUT Y(I)
210   NEXT I
220   IF (X(2) - X(1)) - (X(3) - X(2)) > 0.00001 THEN
      310
230   FOR I = 3 TO 250
240   INPUT C
250   INPUT Y(I)
260   IF C = 9999 THEN 280
270   NEXT I
280 X = X(2) - X(1)
```

```
290   PRINT D$;"CLOSE";P$
300   GOTO 450
310   PRINT D$;"CLOSE";P$
320   PRINT : PRINT "X DATA MUST BE AT EQUALLY SPACED
      INTERVALS ! "
330   GOTO 100
340   PRINT "DATA MUST BE AT EQUALLY-SPACED INTERVALS"

350   PRINT
360   INPUT "ENTER X INTERVAL ";X
370   PRINT
380   PRINT "Y VALUES(MAX NO=250) SHOULD NOW BE ENTERE
      D IN SEQUENCE"
390   PRINT "END WITH 9999"
400   FOR I = 1 TO 250
410   INPUT Y(I)
420   IF Y(I) = 9999 THEN 450
430   NEXT I
440   PRINT "MAX MO OF Y VALUES REACHED "
450 M = I - 1: REM  NO OF POINTS ENTERED
460   PRINT : PRINT "OPTIONS AVAILABLE : ": PRINT "
      (1) STRAIGHT TO NUMERICAL DIFFERENTIATION": PRINT
      "   (2) SMOOTHING FIRST"
470   INPUT "ENTER 1 OR 2 ";Q
480   IF Q = 1 THEN 510
490   IF Q = 2 THEN 550
500   GOTO 460
510   PRINT : PRINT "FURTHER OPTIONS :": PRINT "    (1)
      QUADRATIC POLYNOMIAL": PRINT "    (2)CUBIC POLYNOM
      IAL"
520   INPUT "ENTER 1 OR 2 ";V
530   IF V = 2 THEN 630
540   GOTO 590
550   REM - CONVOLUTING FACTORS FOR 5 POINT SMOOTHING
560 C(1) =  - 3:C(2) = 12:C(3) = 17:C(4) = 12:C(5) =
      - 3
570 N = 35: REM - NORMALISING FACTOR
580   GOTO 660
590   REM - CONVOLUTING FACTORS FOR 5 PT NUMERICAL DIF
      FERENTIATION (QUADRATIC) TABLE III
600 C(1) =  - 2:C(2) =  - 1:C(3) = 0:C(4) = 1:C(5) =
      2
610 N = 10: REM - NORMALISING FACTOR
620   GOTO 660
630   REM - CONVOLUTING FACTORS FOR 5 PT NUMERICAL DIF
      FERENTIATION (CUBIC) TABLE IV
640 C(1) = 1:C(2) =  - 8:C(3) = 0:C(4) = 8:C(5) =  -
      1
650 N = 12: REM - NORMALISING FACTOR
660   REM - READ 5 Y VALUES INTO TEMPORARY STORE
670   PRINT : PRINT "CALCULATING - PLEASE WAIT "
680 K = 0: REM - POINT COUNTER
```

```
690  S = 0: REM - SUMMATION
700  REM - MULTIPLY Y VALUES BY CONVOLUTING FACTORS,S
     UM AND THEN NORMALISE TO GET NEW VALUE OF THE MID
     -POINT
710  FOR J = 1 TO 5
720  D(J) = Y(K + J)
730  D(J) = C(J) * D(J)
740  S = S + D(J)
750  NEXT J
760  IF K > 5 THEN 820
770  REM - STORE FIRST TWO POINTS
780  IF Q <  > 2 THEN 800
790  X = 1: REM  - VALUE OF X INTERVAL NOT REQUIRED IN
     SMOOTHING,SEE LINE 890
800  Z(1) = Y(1)
810  Z(2) = Y(2)
820  Z(K + 3) = S / (X * N)
830  K = K + 1
840  IF K <  = (M - 5) THEN 690
850  Z(M - 1) = Y(M - 1)
860  Z(M) = Y(M)
870  REM - LAST TWO POINTS STORED
880  REM - OUTPUT FOLLOWS
890  PRINT : INPUT "DO YOU WANT A PRINTOUT ? ";J$
900  IF J$ = "N" GOTO 1150
910  PRINT : PRINT "IS PRINTER ON ? Y OR N ? "
920  GET A$
930  IF A$ <  > "Y" THEN 910
940  PR# 1
950  PRINT  CHR$ (137);"80N"
960  IF Q = 2 THEN 1050
970  IF V = 2 THEN 990
980  Z$ = " (QUADRATIC APPROX) ": GOTO 1000
990  Z$ = " (CUBIC APPROX) "
1000  PRINT : PRINT "    FIRST DIFFERENTIALS";Z$
1010  PRINT "=============================================="

1020  PRINT : PRINT "X","Y","DY/DX"
1030  PRINT "----------------------------------------------"

1040  PRINT : GOTO 1110
1050  PRINT : PRINT "          SMOOTHED VALUES"
1060  PRINT "          =================="
1070  PRINT
1080  PRINT "  X","ORIG Y","SMOOTHED Y"
1090  PRINT "----------------------------------------------"

1100  PRINT
1110  FOR L = 1 TO M
1120  PRINT L * X,Y(L),Z(L)
```

```
1130   NEXT L
1140   PR# 0
1150   PRINT
1160   INPUT "DO YOU WANT TO STORE THIS OUTPUT IN A DA
       TAFILE ? ";C$
1170   IF C$ <  > "Y" THEN 1350
1180   PRINT : INPUT "ENTER FILENAME ";M$
1190 D$ =   CHR$ (4)
1200   PRINT D$;"OPEN";M$
1210   PRINT D$;"WRITE";M$
1220   IF Q <  > 2 THEN 1250
1230 B1 = 1:B2 = M
1240   GOTO 1260
1250 B1 = 3:B2 = M - 2
1260   FOR L = B1 TO B2
1270 A = L * X
1280   PRINT A
1290 Z = Z(L) * 10000:Z =   INT (Z):Z(L) = Z / 10000: REM
       DECIMAL PLACES
1300   PRINT Z(L)
1310   NEXT L
1320   PRINT 9999
1330   PRINT 0
1340   PRINT D$;"CLOSE";M$
1350   PRINT : PRINT "FINISHED ? Y OR N ?"
1360   GET B$
1370   IF B$ = "Y" THEN 1440
1380   IF Q <  > 2 THEN 460
1390   FOR L = 1 TO M
1400 Y(L) = Z(L)
1410   NEXT L
1420   REM - SMOOTHED VALUES NOW BECOME INPUT
1430   GOTO 460
1440   END
```

(b) Peak integration

Peak area determination is usually done by numerical integration using either Simpson's rule or the trapezoidal rule. For a large number of closely-spaced points, the trapezoidal rule is adequate and simpler. Limits of integration have to be set after examination of the data using programs FASTDISP or READISP. Note that the X values may be unevenly spaced.

Program TRAPSUM

Input: Filename for data, starting and stopping values of X.
Output: Results on printer, i.e., area at return, area correction, and corrected area as defined in Fig. B.1.

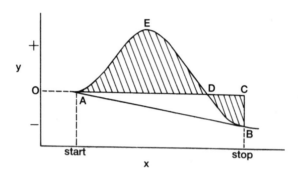

shaded region = area at return = area AEDA − area DCBD
triangle ABC = area correction = area ADBA + area DCBD
corrected area = shaded region + correction
area AEDBA = area AEDA − area DCBD + area ADBA + area DCBD
= area AEDA + area ADBA

Figure B.1 Peak area determination.

```
1   HOME
2   PRINT "**********************************************"
3   PRINT "* PROGRAM TRAPSUM FOR AREA INTEGRATION *"
4   PRINT "*   USING THE TRAPEZOIDAL RULE         *"
5   PRINT "**********************************************"
6   PRINT : PRINT "VERSION AS AT 15/7/85 MEB"
7   PRINT : PRINT
8   PRINT "THIS VERSION IS FOR ": FLASH : PRINT "UNEQU
    AL": NORMAL : PRINT "X INTERVALS ": PRINT
9   PRINT "THE PRINTER MUST BE ON ": PRINT "
      ===========": PRINT
```

```
10   INVERSE : PRINT "DO YOU WANT CUMULATIVE AREA PRIN
     TOUT ? Y OR N ?": INPUT B$: NORMAL
11   IF B$ <  > "Y" THEN 29
12   PR# 1
13   POKE 1657,80
14   PRINT "PEAK INTEGRATION BY TRAPEZOIDAL RULE "
16   PRINT "================================== "
18   PRINT
20   PRINT "POINT  X VALUES  Y VALUES  Y INC        CUM
     ULATIVE AREA "
21   PRINT "--------------------------------------------
     "
25   PR# 0
29   PRINT : PRINT
30   INPUT "ENTER FILENAME CONTAINING DATA  > ";A$
90   PRINT
100   INPUT "ENTER STARTING X    > ";P
110   PRINT
120   INPUT "ENTER STOPPING X    > ";Q
130   PRINT
220 F = 0:S = 0
240 D$ =  CHR$ (4)
249   PRINT D$"OPEN";A$
250   PRINT D$;"READ";A$
260   INPUT X1,Y1
265   PRINT X1,Y1
270 I = 1
280   INPUT X2,Y2
290   PRINT X2,Y2
300 I = I + 1
310   IF X2 = 9999 THEN  PRINT "END OF FILE DETECTED "
     : GOTO 2000
315   IF F = 1 THEN 500
320   IF X2 >  = P THEN 360
355   GOTO 280
360 F = 1:L = Y2:G = I:M = X2
370   GOTO 520
500 T = (Y2 - L) * (X2 - X1)
502 S = S + T
504   IF B$ <  > "Y" THEN 518
505   PR# 1
510   PRINT I;"         ";X2;"          ";Y2;"              ";
     Y2 - L;"                 ";S
515   PR# 0
518   IF X2 >  = Q THEN 600
520 X1 = X2
530 Y1 = Y2
545   GOTO 280
600 S = S - (((Y2 - L) * (X2 - X1)) / 2)
601   REM  END CORRECTION
610 R = (Y2 - L) * (X2 - M) / 2
611   REM  AREA CORRECTION
```

```
615   PR# 1
620   PRINT : PRINT
625   PRINT "OUTPUT FROM TRAPSUM3": PRINT "------------
      ----------": PRINT
630   PRINT "AREA AT RETURN = ";S
640   PRINT "AREA CORRECTION = ";R
650   PRINT "CORRECTED AREA = ";S - R
690   PRINT "===================================="
691   PRINT
692   PRINT "DATA FILENAME   : ";A$
694   PRINT : PRINT "INTEGRATED BETWEEN X = ";P;" AND
      X = ";Q
700   PR# 0
2000   PRINT D$;"CLOSE"
2100   END
```

Explanation of the symbols used in the text

(SEE INDIVIDUAL SECTIONS FOR MORE DETAIL)

A = area; amplitude of vibration; Arrhenius pre-exponential factor; component of binary mixture

B = calibration factor (heat capacity); bulk modulus; sample thickness; component of binary mixture

C = heat capacity; calibration constant; Curie constant

D = thermal diffusivity

E = Young's modulus; emanating power; Arrhenius activation energy

F = force; fraction melted

G = shear modulus

H = enthalpy; magnetic field strength

K = instrument constant; heat transfer coefficient

K_f = crysoscopic constant

L = length

M = magnetization; molar mass

P = period of oscillation

Q = heat

R = gas constant; thermal resistance; recorder sensitivity

S = entropy; chart speed; area

T = temperature

V = volume

Y = general ordinate

a = constant; partial area; activity

b = constant; temperature integral (section 13.5)

c = specific heat capacity

d = diameter

e = exponential base
$f()$ = function
$g()$ = function
(g)   = gas
$h()$ = function
h = baseline displacement; height; relative amplitude of vibration
k = rate coefficient
(l)   = liquid
m = mass; constant exponent in rate equation
n = apparent order of reaction; number of moles
p = pressure; constant exponent in rate equation
$p(x)$ = integral defined in section 13.5 ($x = E/RT$)
q = heat
s = partial area
(s)   = solid
t = time
x = mole fraction; dimension
x_1 = mole fraction of solvent
x_2 = mole fraction of solute
x_2^* = make fraction of impurity
y = dimension
z = dimension

Δ = change or finite difference; damping
α = coefficient of linear expansion; fractional reaction
β = see chapter 4
ε = strain; dielectric constant; undetermined premelting
η = viscosity coefficient
θ = angle; Weiss constant
$\theta(T)$ = function of temperature
κ = volume magnetic susceptibility
λ = thermal conductivity
μ = chemical potential
ν_p = Poisson's ratio
ν_r = resonance frequency
ρ = density
σ = stress
ϕ = heating rate
χ = mass magnetic susceptibility

Subscripts
c   = calibrant
f   = furnace; fusion

g   = glass transition
gm   = geometric mean
m   = melting
o   = initial
p   = constant pressure
r   = reference
s   = sample

Index